Doblin
Feiner
Finzer
Hooper
Hulteen
Marcus
Mauro
Owen
Whitney

Design in the Information Environment

How computing is changing the problems, processes and theories of design

Editor: Patrick Whitney
Institute of Design
Illinois Institute of Technology,
with
Cheryl Kent

Southern Illinois University Press
Carbondale and Edwardsville

First Edition

987654321

 Published in the United States by Alfred A. Knopf, Inc., New York, and simultaneously in Canada by Random House of Canada Limited, Toronto. Distributed by Random House, Inc., New York.

ISBN 0-8093-1251-4

Manufactured in the United States of America

Preface

This book is intended to be an introduction to the new forces influencing the various fields of design as we move from the industrial age to the information age. The articles are based on presentations at two conferences held at the Minneapolis College of Art and Design. The contributors were selected from various fields including visual communications, product design, architecture, computer science, and psychology. The diversity of the people is indicative of the changing environment in which designers find themselves: the answers to our questions about changes do not lie in a single body of knowledge. The projects and ideas in these articles represent exciting directions that are important to designers interested in the future.

Of the many people who helped with both the conference and this book, a few require special thanks: Richard Mueller and Peter Seitz for giving early support and participating in the conferences, Gisela Erikson and Suzanne Richards for working on the production and design, and the Dayton-Hudson Foundation for providing initial funding.

Patrick Whitney
Institute of Design, IIT

Table of Contents

1 **Foreword**
Katherine McCoy

8 **Information, Computers and Design**
Patrick Whitney

18 **Information and Design: The Essential Relation**
Jay Doblin

32 **Changing Pictures/Changing Minds**
Kristina Hooper

46 **The Know Business is Show Business: Graphic Design and Computer Design**
Aaron Marcus

56 **The Interface to Design**
Eric Hulteen

66 **Coping with Complexity**
Charles Mauro

82 **The Department of Crude Arts: Viewing Videotex and Teletext from the Graphic Designer's Perspective**
Aaron Marcus

94 **Computer Graphics and Visual Literacy**
Charles Owen

118 **Interactive Documents**
Steven Feiner

134 **Programming by Rehearsal: A Graphic Programming Language for Computer-Based Curriculum Design**
William Finzer

154 **Computers in Design Education**
Charles Owen

168 **Biographies and Reading Lists**

Design in the Information Environment

Foreword

The Industrial Revolution ended in 1950, yet we are only now realizing that the information environment is upon us. We see now that it will be up to us, as designers, to determine our role in this new age. If we do not, others will. Our alternative is our own obsolescence. Industrial designers and graphic designers, in particular, could soon be remembered with the same affectionate nostalgia as blacksmiths and linotype men.

As part of this post-industrial information society, heavy industry is declining and the services increasingly dominate the manufacturing sectors in our economy. Some industrial designers are alarmed to observe the service industries replacing the belching smokestacks of manufacturing. Since industrial designers design mass-produced goods, they wonder if they will have a role in the future, if the trend is really away from our industrial base. Industrial design would seem to be intrinsically tied to the industrial revolution. The answer to this question lies in the direction the field pursues in the next few years.

Graphic designers produce visual communications, on the other hand, and have always dealt directly with communication, their central purpose. We are witnessing a communications revolution, an explosion of available information; or if not information, at least data. The challenge to graphic designers will be to give clarity, order and meaning to this deluge of electronic data, as Hulteen suggests later in this book.

The design of environments will also incorporate the information factor. Offices, in particular, are information processing centers or exobrains, as Doblin calls them. A major result of the post industrial reordering is the mammoth shift in the workplace, from factory to office. Interior designers and architects will progressively find communications and information flow prime determinents of their design solutions. But no environment will escape the revolution. Homes, schools, stores, museums, amusement parks, and factories are all transforming into electronic media experiences.

Both the graphic and industrial designer, in particular, have vast new areas open to them. The very materials of our practice are changing, the result of the computer. The current interest (and panic in some sectors) in CAD/CAM is the tip of the iceberg.

Paper is turning into electronic impulses and fleeting images on CRT's; plastic and metal are becoming software and microchips. Print communications and products are dematerializing, becoming less tangible. The output of our design activity has been familiar and comforting hardware up to now, part of the long history of artifacts the human race has grown up on. Now these artifacts could ultimately become the dematerialized delivery of information and services. (Of course, this has always been the final purpose of visual communications and products.)

Ettore Sottsass predicted this with his service grid landscapes in his utopian proposals of the late '60's. Short of this, even now we see products miniaturizing radically with the advance of technology. Others are integrating into systems, blending with other previously discrete products. Mechanical parts are disappearing. Without bulky moving parts to determine form, many products are becoming black boxes, existing mainly for the display of operational controls and marketing messages to the users.

The question here is, how shall we design for non-artifact design solutions? Tangible artifacts are what we've been trained to produce, what our design school educators know how to teach, and what consumers over age twenty are used to owning. And the human race, after all, has been artifact-oriented for perhaps two million years.

In anticipating such a radical impact of the information revolution, we must consider the alternate possibility that perhaps not all our output will dematerialize. If one is to believe the high tech—high touch principle of compensating balance (quite possibly more than just a catch phase), our audience and users will always need some reassuringly tangible hardware as a link to our past genetic history and to the physical world. We may all require the physicality of a finely bound book or a ritual artifact like an ornamental espresso machine to counterbalance the highly cerebral world of the information environment.

Designers now seem ready to accept the idea of computers as a tool, replacing hand labor in the repetitive aspects of the design process. This is generally the documentation phase, what Pat Whitney terms 'forming the solution': architectural specifications, ready-for-camera artwork, shop drawings, interior furnishings orders, business records. The economic savings of computerizing such labor-intensive activities is attractive and is the threshold level of the computer revolution for most design offices.

But beyond this, where do we go? Certainly there is a universe of possibilities, and this book begins the process of selecting options for development.

A 1983 article in Esquire magazine on the future of the engineering profession had some interesting points that relate to our future as industrial and graphic designers. The article said, "a post-industrial society requires more than just a lot of engineers. It requires engineers with a vision and foresight to see where they are going and the verbal capacity to tell us about it. But American engineers are taught the value of analysis, of pulling apart, but not of synthesis, of putting together. This raises the question of whether America is philosophically suited to leadership in the post-industrial world—of whether this country would in some basic way may be incompatible with the future."

The design professions have some answers to the problems that engineering faces in this new order. Designers are specialists at synthesis. The new world of high technology needs more than analytical engineers, and designers can provide that necessary input, making the connections.

Designers now have the choice to let engineers shape electronics (for even in this new world, information and services must take some form, even if they are no more finite than the appearance of a video screen display or the voice quality of synthetic computer speech). Or designers can shape these new electronic tools themselves.

Engineers are now designing the computer hardware and software necessary for solid modeling, spatial simulation and computer animation. They are also designing typefaces, screen formats and business graphics formats software. And we are presently letting engineers conceive the "smart products."

But designers are the experts in these areas. We must master the new media and become central to the new production process. We are entering a communications age and the key role of industrial design activity will involve communication-communications between man and machine. How to operate machines. How to understand them. Products must communicate their effective, safe and satisfying operation to their users; users must communicate the services required to the product. Many machines themselves will be primarily communication devices. The electronic media will be involved in many different areas, in addition to computers, office communications, telecommunications and home entertainment.

The industrial designer's input may be largely in that area described as high touch. High touch will be a very important compensating balance to the abstraction of the high technology that is shaping our daily experiences. Industrial designers are the specialists at this, the experts on the man-machine interface.

Designers must position themselves as the experts at humanizing the machine for their users. Human factors and ergonomics are presentlv interpreted as the physical interface. But the concept of human factors must be expanded to include the psychological, cognitive, and perceptual interface with the user. And designers can appeal to the senses, using their aesthetic abilities to add a sensual aspect to the machine that will allow users to participate in technology more comfortably. Technology can and must be brought to life, animated and humanized. It is the designer that has the vision to demystify technology and warm up the abstractions of a digitized world.

We are entering the era of smart products. Many new features are being introduced into existing products using microprocessor technology. Whole new product types are emerging in an increasing avalanche of possibilities in the area of communications and information processing for business, professional and consumer markets. The synthesis of smart product concepts involves both sophisticated engineering and marketing, two fields that do not interface well. This is precisely where the industrial designer's link is essential. Designers are in a unique position, able to draw from both fields and add an understanding of the human response.

Industrial designers are the experts to initiate original product concepts, the features they offer the consumer, and the way the user interacts with them. A case could be made for a corporate super-department of industrial design that would house all aspects of the product development process: advanced product planning, engineering research and development, product design and marketing.

For graphic design, electronic technology will also be more than a design tool. Electronic delivery systems are changing the physical output of visual communications in graphic design, illustration, signage, and exhibition design. New media forms are developing daily, it seems, including viewdata and videotext in electronic publishing. In addition there are digitized typefaces, screen formats, computer interfaces, and business graphics formats to be designed. If graphic designers position themselves correctly, their field could become larger.

A major change in the graphic designer's sensibility could result from the addition of movement to the graphic designer's vocabulary of presently static design elements. Animated typography and images, now only incorporated in television graphics and film titles, will replace the idea of page sequence in graphic design for video display.

The distinction between industrial design and graphic design will blur, as products become more communications and less mechanical function. Already many products are only as large as the control surfaces require to operate them. (Wristwatch calculators, for instance, could be far smaller but must be big enough for data entry by human fingers.) The complexity of programmable multi-featured products requires all the skills of visual communications to render their operation clear and comprehensible. Presently industrial designers are generally responsible for "product graphics," but soon hybrid designers may specialize in this area of product communications. The dominant visual form of many products may become mainly graphic-verbal, numerical and symbolic.

As to the larger view of the information revolution, the possibilities are almost overwhelming and the degree of abstraction is unsettling for traditional designers who are very much into the sensual tangible world around them. Designers traditionally have been the expert manipulators of physical phenomena in a very hands-on manner. One questions how well we can deal with this new abstraction that translates the tangible to the numerical. What happens to the hand-eye connection intrinsic to the idea generation process when designers 'draw' on video screens?

Of all the changes, the most staggering is the volume of available information. Everyone, and especially the designer, faces the task of translating data into messages of meaning and information into wisdom. The challenge will be to turn capability into sensibility.

Katherine McCoy

Co-chairman, Design Department, Cranbrook Academy of Art
President, Industrial Designers Society of America

Information, Computers and Design

Patrick Whitney

Most of the discussion about the use of computers in design is focused on computer graphics, and computer-aided design and manufacturing. Although using computers as an aid in production will be useful, the main aid to design will be helping in the analytical and planning processes. These will help designers decide what problems need to be solved as we moved from an industrial society based on material and products, to an information society based on messages and services.

The Dilemma of the Specific and the General

In the Yucatán peninsula corn is planted by Indian farmers in the same way it was done hundreds of years ago. The farmer wears a sack filled with seed slung over one shoulder. As he walks the field's rows, he uses a long stick to make holes in the ground into which he drops seeds. Although the stick is a simple tool, it is not naive. It has features that make it well-suited for its task: it is long enough so the farmer can make the hole without bending to the ground; and, the end of the stick is sharpened to a point to make the hole for the seed.

In some developing countries, one can still see people, like the farmer in the Yucatan, who are both the users and designers of their tools. In the industrialized world, the roles of designer and user have become separated. Though separate, the user-designer and the modern designer should share certain obvious concerns. It has always been the business of design to affect human experience and behavior by shaping elements of the physical world. Human artifacts such as buildings, cities, agricultural implements, industrial tools, books, and signs, are but a few examples of the items we make to support human survival, safety, comfort, communication, pleasure and rituals.

Craftsmen were, in a sense, precursors to today's designers. Craftsmen specialized in making needed goods for members of their immediate community. Unlike contemporary designers however; craftsmen physically made products; they developed expertise in manipulating tools and materials; and the things they produced both followed and reinforced the traditions and values of the communities in which they lived.

It was not until the Industrial Revolution that a complete separation between the tool-maker and the tool-user occurred. Rather than hand-producing everything for a particular person who requested it, the designer was paid by a manufacturer, developer or publisher, to specify the characteristics of artifacts made on equipment meant to make large quantities of identical products. The things he designed were for people he would never see. Not only was he separated from the users, he found himself writing and drawing specifications that would allow someone else to actually produce the item.

With economy of scale, the number of items produced were increased and individual consumers were transformed to become "the market" which grew continually larger. Producers found it economically advantageous to design for the average characteristics of a large number of users. This led to relatively low costs, and a trend for products to be *just good enough* to meet a wide number of needs. This meant in turn, that no mass-produced product could exactly match the needs or wants of a single person or group. Designing for a mass audience required a large number of compromises to be made in order to reach the level of average acceptability.

This has been one of the major dilemmas of design in the industrial age: how can one meet the needs of individuals when the central characteristics of mass production lead to messages, artifacts and buildings that are made for the average?

There is some differentiation in manufactured products. Of necessity, products come in different sizes; and, in any one general product type, there is variation in price. These variations match roughly the needs and wants of different segments of the population. But clearly, they are no more than crude gestures when contrasted to the "custom-made" work done by craftsmen for individual customers. The designer is left in the ironic position of professing concern for the user, while being restrained by mass production tools that allow him to design only for the average.

Of course, it is the economy of large production runs and other characteristics of industrialized society that has helped the Western middle class achieve its materially abundant life. The computer's capability to deal with specific details within a large amount of information allows one to consider producing more specialized artifacts.

Computers, communication systems, a highly educated population, and other contemporary developments are bringing about major structural and social changes. We are now at the beginning of an information revolution which is characterized by two major changes: the availability of a new *design tool* and a population living in a new *information environment.*

New Design Tool

The first change—one that speaks directly to the design dilemma—is the use of the computer as a design tool. Interactive design stations allow designers to explore far more alternatives than have been possible with traditional tools. Robotics and cyber-automated production processes are allowing both the economy of scale and a wide degree of variation in individual artifacts. Numerically-controlled printing presses and production equipment are far more flexible than offset printing plates and metal production dies. The new flexibility permits alterations to be made in a product or publication according to the needs of an individual or a group. Changes are made by manipulating the information encoded in a computer database, which in turn, instructs and guides the machinery making the product.

In less technologically advanced production processes, information cannot be manipulated so simply; the information is not stored in a database, it is frozen on an offset printing plate, or in a metal production tool. As a result of the new technology, components for houses, products and printed pages can be altered to meet specific needs rather than average needs, while still being cost effective.

It is paradoxical that computer-supported design and a highly automated production process can restore the specialized design quality that was taken away by the Industrial Revolution. These are, after all, qualities we associate with phrases like "hand-made", and images of solitary craftsmen working carefully, laboriously, to perfectly execute every detail of their product.

To the designer, this design/production process offers the chance to design for the user in a way that has not been possible with mass production. But, for the time being, this remains only a potential use of the computer. For this much-acclaimed new tool to achieve this potential requires the design process to take into account the needs of specific users. What is needed then, is an effective way to gather and organize information in a manner that helps lead to more intelligent solutions to problems.

Of course computers are ideally suited to helping designers analyze, organize and evaluate information. The analytical aspects of the design process, however, seem to be forgotten in all of the discussions about computer-supported design. It is here that the computer has a critical role to play. If one looks at the general divisions of the design process (defining the problem, research, idea development, forming the solution, production and evaluation), computer applications can be found to support each one.

Defining the Problem (analysis)

In complex problems such as the development of new lines of products, buildings, signage systems for complex facilities, or corporate identity programs, it is difficult to keep all of the factors of the problem in an order that allows one to see all of the relationships. There are computer programs that allow one to list the attributes of the various parts of a problem and restructure them into new relationships that lead to new insights about the basic problem.

Some of the most advanced work in developing computer-supported methods for structuring problems is being done by Professor Charles Owen at the Institute of Design, Illinois

Institute of Technology. His computer programs are designed to break complicated problems into elements which can be reorganized into various structures. The designer evaluates the structures for appropriate and insightful relationships.

Research (analysis)
If doing research for a design problem includes gathering information and reorganizing it into useful components, then there are two main ways the computer can be used in this phase.

Gathering information can be helped by computer-assisted searches through banks of information. At its simplest level, this takes place through bibliographic searches directed by key words. As information about design accumulates and the cost of using these systems comes down it is likely that a computer information network dedicated to information about design will develop. The computer can also be used in a more basic information gathering mode called data capture. As its name implies, this technique does not gather information, but data, which through analysis is turned into information. Data capturing is used to monitor environments to develop information related to human factors such as temperature, humidity, sunlight, and noise levels. The same procedures can be used to monitor physiological characteristics such as muscle fatigue, heart and breathing rates, and eye motion. The quickness, patience, and objectivity of the computer make it a very desirable monitoring system.

Its usefulness is enhanced if the computer can also organize data into a meaningful hierarchy. By taking information from networks and data from the environment the computer can organize it into structures that relate to a problem.

Developing Alternative Ideas (synthesis)
This part of the design process is often characterised by drawing, building models, or trying alternate text/image relationships for printed materials. In this phase, the computer is

used to develop two-dimensional images of these alternative quickly. As with the computer industry in general, the cost of systems is coming down as the capabilities are increasing. As this leads to affordable systems that display relatively high quality images, it allows the designer to rapidly make, view, modify and review images. *It alters the process to one that combines the versatility of drawing and the speed of collaging.* The sketching done on the machine produces images which can then be evaluated. Because the drawing can be done more rapidly, more ideas can be tried, presumably resulting in better, more appropriate solutions. This process is being extended into the design and production of products by connecting the database used for design to numerically-controlled milling machines which produce models and molds more efficiently than those done by hand.

A more current aid to the development of three-dimensional objects is solid modeling. This is a process in which an image can be displayed in three-dimensional perspective. Although these images do not look significantly different from perspective surface drawings, they are significantly different in that they contain information about the material proposed for the product. The product can be tested for impact, wear, and user requirements by computer processes before it is physically made.

Forming the Solution (synthesis)
The same functions that allow one to view alternate images can be used to produce working drawings, floor plans, elevations, typeset text, layouts, and enhanced photographs. The complex plans (figure 4) are a good example of this.

Production (synthesis)
Robots, laser printing and numerically-controlled manufacturing equipment can be guided in production by the use of the same database as that used in design development. This potential could eliminate a good deal of what lies in the middle area between design conception and design production. (This includes

the people producing printing mechanicals, technical drawings and plans, and assembly line workers. Conversely, it also offers the potential of new design activities, particularly in the analytical areas.)

Evaluation (analysis)
The process of evaluating a message, product or space can be done by using systems similar to those used in the research phase. These would gather data and organize it into a hierarchy relevant not only to the design solution at hand, but to the entire field of design. The information gathered in such analysis can be used in building a body of knowledge about design issues that can be used in the future.

Information Environment
The second major change is that members of our society are living in an information environment. Evidence of the change is plentiful; for example, there has been an increase in the percentage of workers in knowledge industries and a decrease in manufacturing, mining and agriculture.

The shift in occupation is evidence of other broad changes occurring in society that have consequences for design. The type of work that people do has changed, their level of education has changed; and, not surprisingly, their tastes and buying habits have changed as well. There is evidence, for example, that some consumers are willing to buy expensive items of very high quality while scrimping elsewhere. A consumer like this might have an $800.00 Italian espresso machine and serve espresso in cracked mugs. Our hypothetical consumer may be isolated in his extreme devotion to espresso, but he is not alone in having a highly sophisticated taste for a particular product. What is becoming increasingly clear about the information environment society is that its members have very discrete tastes and are segmented into small groups. Predicting the tastes and desires of these consumers is difficult, if not impossible. To design for today's society one needs to have an understanding of the new patterns, and an understanding of the means of designing for it.

While today's society is radically different and more complex than yesterday's, we now have the quintessential tool for helping us respond to complexity.

With few exceptions, the use of computers in design has been limited to supporting the synthesizing phases of the design process. This is somewhat ironic because the computer gives us the ability to rapidly gather and analyze information; and the information environment is presenting us with a situation as complex and rapidly changing as mankind has ever faced.

The inherent advantage in using the computer as a tool is in increasing speed, not quality. The assumption that quality will follow is based on the assumption that the people producing the larger number of images will have the ability to evaluate them and make a choice about which is best to use. The use of this new equipment will certainly not be limited to designers and others with critical facilities. What is likely to happen is that in the hands of a good designer, computer graphics systems will allow him or her to do more good work. In the hands of others it will mean more visual garbage than the world has ever seen.

The central question facing the design field is not simply how to use these computer tools to produce solutions more rapidly; but to use computers and to gain a clearer understanding of what should be designed to fit the new context of the information environment.

Information and Design: The Essential Relation

Jay Doblin

With the digitizing of information, human beings are endowed with the powers once ascribed to gods. Jay Doblin reflects on the nature of information and the impact of its digitization.

I am going to theorize about information - something as perilous as dancing on quicksand.

Information in Design

Every product is composed of information, materials, and energy. The form given to the materials is the result of applied information, and the materials are formed by an expenditure of energy. For example, the materials in an automobile are some dirt and oil. By successive applications of information and energy, a car is produced. At the end of the assembly line that car reaches its peak of integration.

The process by which products are made can be described by a flow chart with five components: **control, operation, product, consumer,** and **evaluation.** (Figures one and two.) The process begins in **control** with the definition of the project's purpose. Information generated in **control** is sent to **operation** where it is combined with materials. The result is the product.

That product is evaluated twice, objectively and subjectively. In the objective evaluation, direct measurements are taken from the product: its size, weight, durability, and capacity. Subjective evaluations are indirect; they are based on consumers' opinions and preferences. Eventually, both the subjective and objective data is sent back to control. This flow model is a cybernetics loop. Information from evaluation is the feedback used by control to make corrections and refinements in succeeding cycles.

What flows throughout this system is information. The arrows are information transmission, the cells are information transformation. To understand this model properly requires an accurate definition of information.

Information in Products

Of the thousands of cases where information is used for

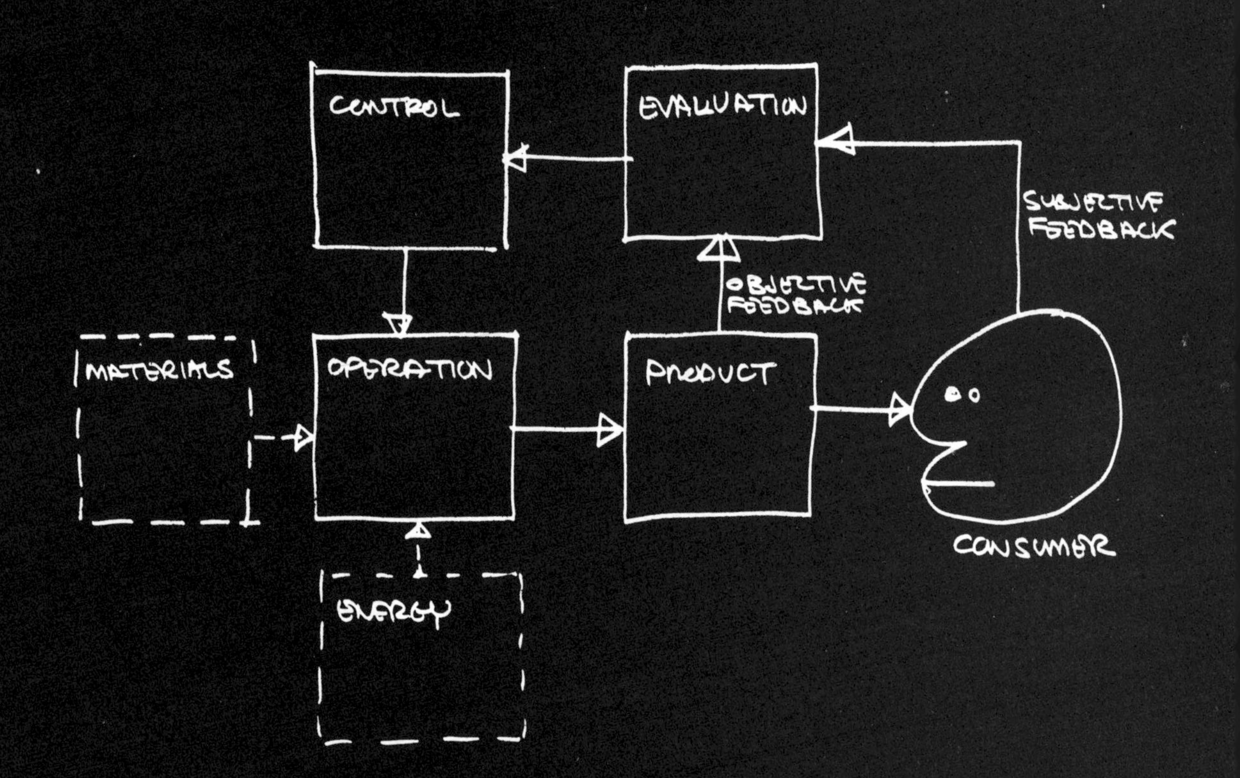
CONTROL
EVALUATION
SUBJECTIVE FEEDBACK
OBJECTIVE FEEDBACK
MATERIALS
OPERATION
PRODUCT
CONSUMER
ENERGY

Figure one.

The Process of Making a Product is outlined in Jay Doblin's five step flow chart. (1) In control, what the product will do, and how it will work is determined. (2) From control the plans go to operation. Here, the product is made in accordance with the information from control, and in combination with the appropriate materials, and necessary energy. (3) When the product is complete, it can be objectively measured for size, dimension, strength, and so on. (4) The consumer who has bought the product is also evaluating it, perhaps using different criteria. The product may work, but perhaps it's too difficult to make work, or it's unattractive. These subjective judgments are combined with the objective feedback in the evaluation step. (5) The evaluation is sent back to control where modifications to improve the product can be made.

control, the ones that interest designers are those where information is used to give performance and appearance to products. A product is concentrated information. A Polaroid SX-70 has a massive amount of accumulated mechanical, optical, electronic, and chemical information packed into it. This is what makes it possible for an amateur to produce color photographs. (Consider the skill and equipment it took fifty years ago to produce color photographs.) What happens is that as products evolve (accumulate information) they get smarter, and this smartness reduces the need for smartness and experience on the part of the user.

The process can be reversed, so that information can be taken from the object. Looking backward, an archaeologist can "read" an artifact because a product is "frozen" information, the totality of everything its maker knew about that product at the time it was produced.

Information in Inorganic and Organic Objects

The information that controls the form of products comes third in a long chain of events. The first event, non-informational, occurred when materials and energy interacted to produce natural objects. Examples at the cosmic scale are the formation of suns and planets; at earth's scale, the formation of clouds and mountains; at atomic scale, the formation of atoms and molecules. Science tells us that no information controlled the formation of these objects; they happened. (But such objects can be deciphered into information; this is the major activity of physical sciences.) Page one of the Bible describes God supplying the information to form earth and its objects.

The second event, the spontaneous beginning of natural information, happened billions of years ago when life first appeared on this planet. Science tells us that the first organism was the chance result of some primordial event. Since then, the evolution of species is the result of the natural accumulation

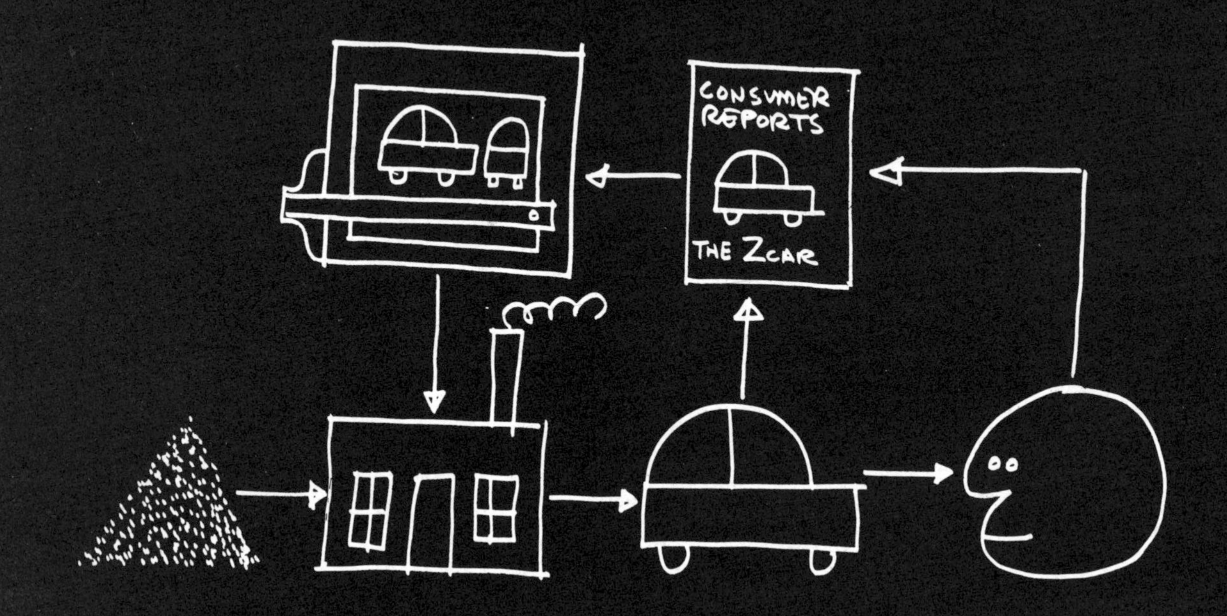
CONSUMER REPORTS
THE Z CAR

Figure two.

The Process of Making a Car, using the same steps described in figure one.

and transmission of genetic information controlling the form of each succeeding generation.

Increasing Diversity and Complexity of Organic Objects

The word **form** appears in the middle of the word **information** and is defined in Webster's Dictionary as, "the shape or structure of something as distinguished from its material". The primary definition of the verb **inform** is, "to give material form to". It is genetic information that gives form to matter to produce an organism. During successive reproduction cycles, favorable genetic information accumulates, and unfavorable information is purged. This causes living things to evolve into more refined, diverse, and complex forms. Charles Darwin wrote in 1859 in his **On the Origin of Species**: "As many more individuals of each species are born than can possibly survive, and as, there is a frequently recurring struggle for existence, it follows that any being, if it varies however slightly in any manner profitable to itself, under the complex and sometimes varying conditions of life, will have a better chance of surviving, and thus be naturally selected. From the strong principle of inheritance, any selected variety will tend to propagate its new and modified form."

Information in Artificial Objects

The process of evolution through the transmission of information seems to be true of products as well. As with organic species, every product can be traced back through its evolutionary stages. From the original flint cutting tool, we now have hundreds of diverse tools: butter knives, engine lathes, bayonets, scalpels, chain saws, guillotines, chisels, safety razors, and so on. As with living things, every product is the accumulation of information, a continuous strain that began with the first primitive tool. But where information in living things accrues genetically and is passed on during reproduction, products do not reproduce themselves (no one has grown a Volkswagen by planting a bolt, or by mating a female

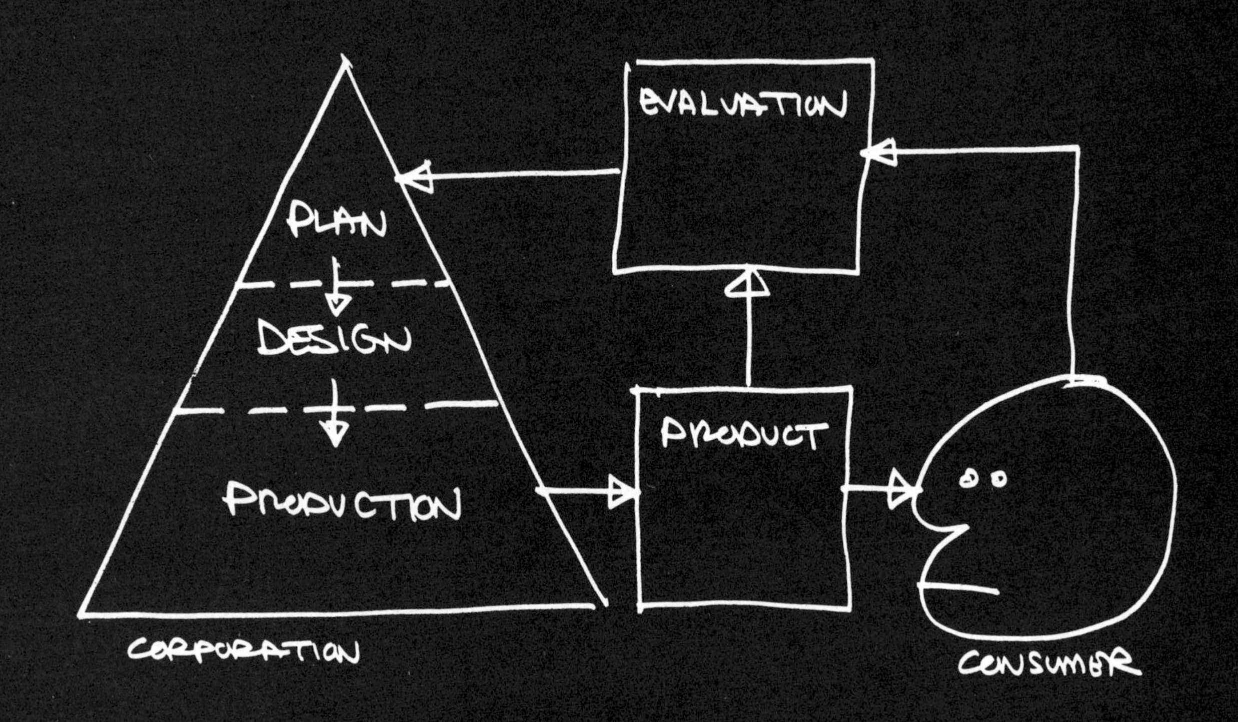
EVALUATION
PLAN
DESIGN
PRODUCTION
PRODUCT
CORPORATION
CONSUMER

Figure three.

Offices as Information Processors (1) In planning, what will be made by the company is decided upon. (2) In design, the planners' idea is translated into practical plans detailing how the product will be made. (3) It is made in production according to design's specifications. (4) The product is evaluated objectively by scientific measurement, and subjectively by the consumer. (5) The result of that feedback is an evaluation which passes back to the corporate planning level and inspires changes to be made in the product.

Volkswagen with a male Volkswagen.

Like organisms, products as they evolve accumulate information and become more complex. But where organic species evolve slowly, products evolve rapidly. The reason is that the information in products can be externalized; designers can get it, extract it, and change it to fit their objectives (bio-engineers now claim that this can be done with organisms too). The information extracted from a product is the de-materialized product. There is one big change in this process—it has shifted from analogical to digital.

Analog to Digitized Information

For millennia, information has been analogical, communicated as words, numbers, and drawings. In this form, information is perceptible to the eyes or ears, and so is easy to use. But it is also cumbersome. In the 1950's, main-frame computers made it possible to use digital information, and suddenly an entirely new capacity was available.

Digitizing makes some extraordinary things possible. Because of digitizing we can collect and store a vast amount of information in a small space, (eventually you will be able to go to the drug store, buy the Library of Congress and carry it home under your arm). With digitizing, we have the possibility of instant access to any piece of stored information. With the right software, that information can be processed in any way we like, instantly. Digitizing also allows the transmission of information to, or from, any part of the world instantly. And, digitizing permits us to display information as copy, lists, charts, or any other desired form. With this power available, all of us become omniscient.

During the last decade, the development of the micro-processor has allowed this power to become compact, economical, and available. We are only at the beginning; how this technology will be used for design is still evolving. At present, these methods are often as frustrating as they are useful. But knowing something about this technology will prove to be critical. We are in a transition period where the haves - who can use this power - will be separated from the have-nots who can't.

Figure four.

Traditional and Electronic Offices
In the traditional office, information is constantly filtered through a hierarchical system. In the electronic office, the hierarchy still exists, but the information contained in the computer is accessible to everyone. This allows a radical change in office problem solving. Groups of people may, under this system, work on a single problem at the same time.

Information in Offices

The term **office** is as vague as **information.** Because there are millions of offices, and because they vary enormously in size and purpose, it is difficult to generalize about them. But whether an office has a staff of one or thousands, whether it is a bank or a newspaper, one thing can be said about all of them - they are information processing operations.

The product model shown earlier can be adapted to demonstrate how offices process information. **Control** and **operation** can be combined here in a triangle that represents the three levels of office operations: **planning**, **design**, and **production**. (Figure three.) Planners, at the top, decide what the company will make. In design, those concerned with human factors, engineering, industrial design, and graphic design, turn the plans into precise specifications that detail everything from color and material, to the last bolt. In production, the instructions from design are followed to create the final product. In office parlance, these three levels are called executive, professional, and clerical.

The Office as Exobrain

The shift from analogical to digital information is causing enormous changes to occur in offices. In the evolution from the traditional office to the modern office, the equipment has changed from that based on analogical message communication, to machinery that will translate the analogical to digital. We have moved from the quill pen, to typewriters, to telephones, to the copier; and into word processors, terminals, micro-computers, electronic typewriters, and laser printers.

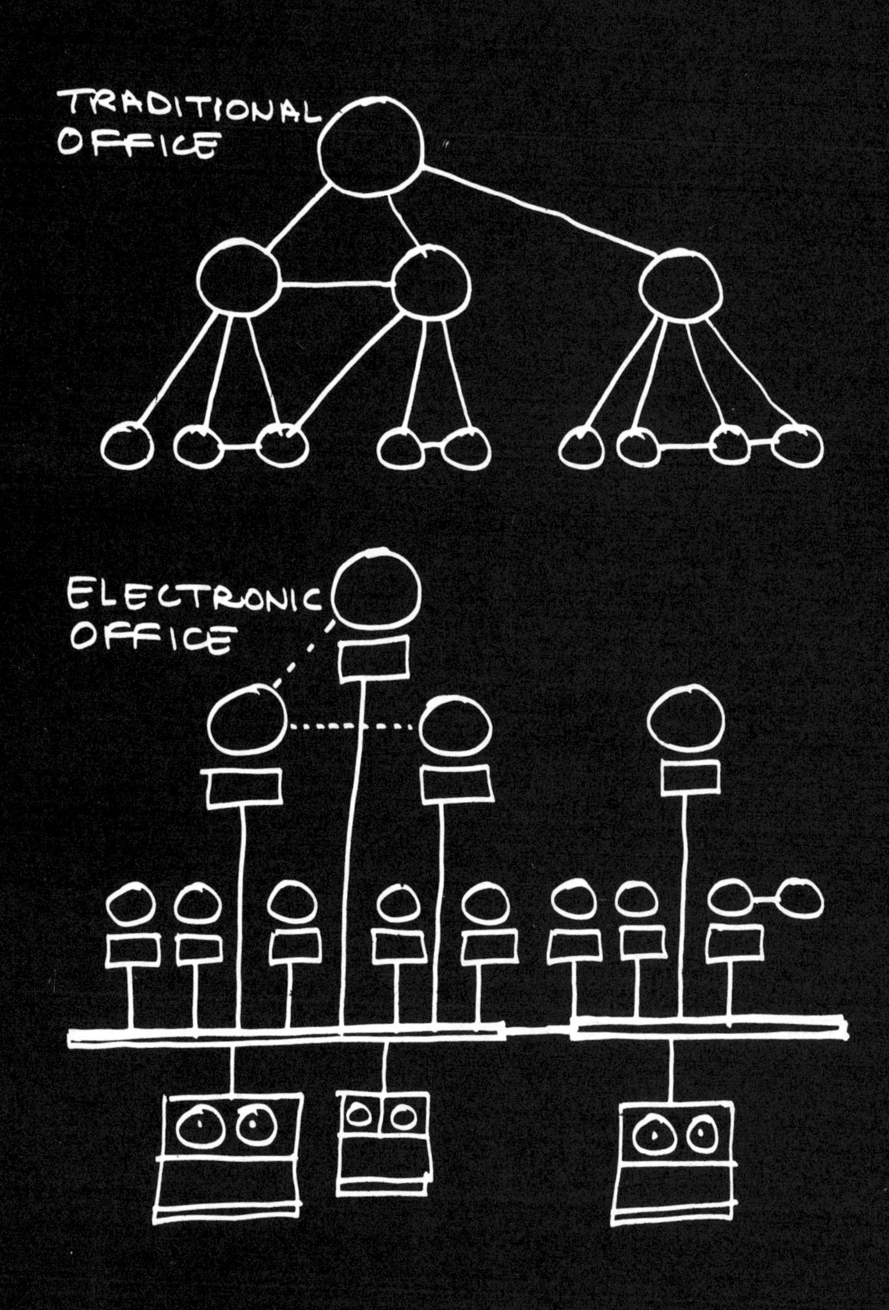
TRADITIONAL
OFFICE
ELECTRONIC
OFFICE

What this means is that many workers can simultaneously work on a single problem because they are tied together through a computer and computer terminals. (Figure four.) The office becomes an "exobrain" in which workers are contributors to an ongoing cumulative process of combined thinking. The result will be a competition in which Mobil's exobrain is pitted against the exobrains of Shell, Texaco, and Standard. May the best exobrain win (that is, after all, the ultimate purpose of an office).

Figure five

Inexorable Trends

Inexorable Trends

There are some inexorable trends that cannot be ignored. The first is the massive shift to office work. At the turn of the century, over eighty per cent of all workers were in agriculture; by the end of this century, two per cent will be. The number of industrial workers grew from a small percentage to over fifty per cent by mid-century; now this number is declining. It is said that it will take 15 per cent of the work force to produce all the food and products we can use. Information workers are increasing rapidly, to over 65 per cent. The major shift is from material-energy workers to information workers.

The second inexorable trend is to digital information. By the turn of this century, most people will be using more digital than analog information. This will not only change our social interaction, but will change how we think. People born into a digitized world will think very differently from the way we do.

To be digitized requires equipment that couples the workers and the system. To be effective, or just employed, every worker will need to be attached. This explains the exponential growth of office equipment - said to be thirty per cent each year.

But I urge you to ignore, if you can, the equipment. The important thing is to understand that these individuals will be combining their skills in a cooperative system. For designers,

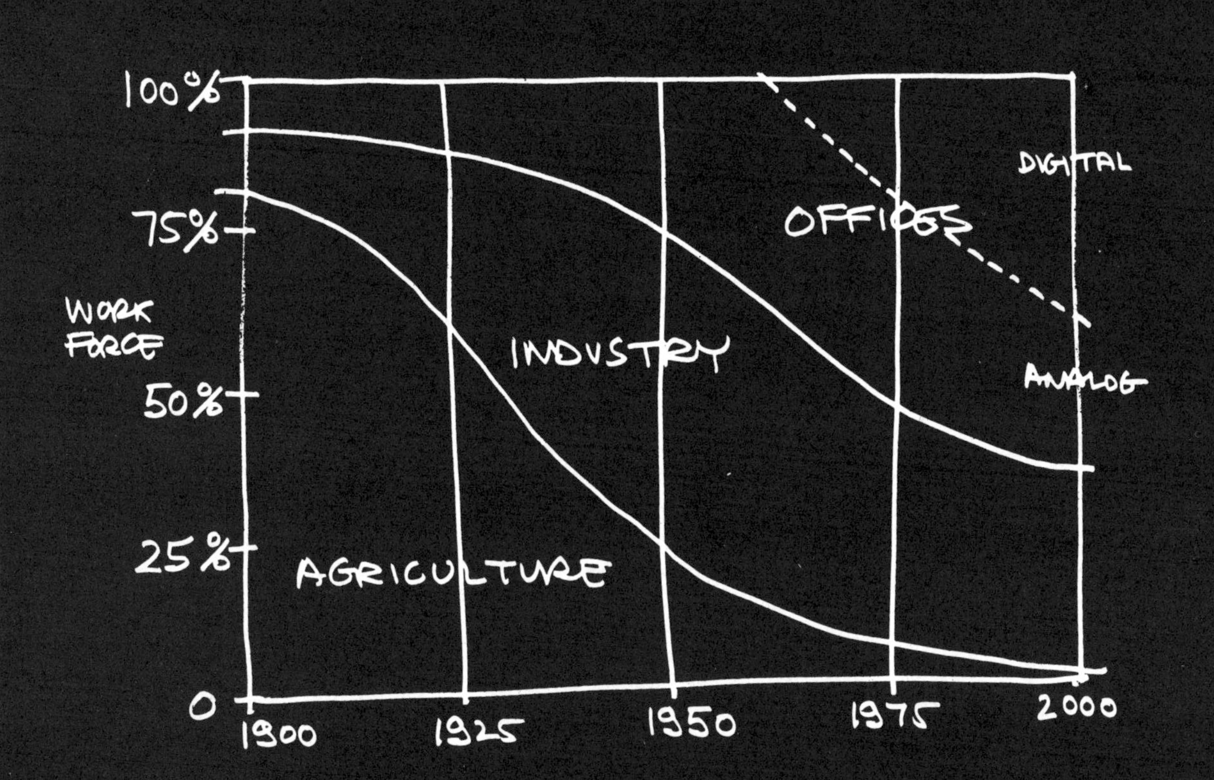
100%
75%
50%
25%
0
WORK FORCE
1900
1925
1950
1975
2000
AGRICULTURE
INDUSTRY
OFFICES
DIGITAL
ANALOG

this means a new way of working that makes obsolete the old board artist concept.

The New Designer

A few decades ago, designers were the visionary stars of business. But now, consulting offices are giving way to corporate design departments where designers are "matrixed" into teams with engineering, marketing, production, and the rest. In the process, as designers homogenize and "professionalize", they may give up their prime purpose: to be corporate visionaries. Designers should be wary of becoming digitized participants who make their specialized contribution. They must keep their most valuable asset - their vision - and use that vision on behalf of the end user, not the intermediate market, or the immediate corporation. This may sound idealistic, but it's not; it's hard-headed good business.

Changing Pictures/Changing Minds

Kristina Hooper

Computers can be thought of as sophisticated production aids. But the computer's dynamic, interactive qualities also make it ideally suited to serve at a much higher level, as a helper in the process of developing ideas. In this capacity, the computer provides a setting for thoughtfulness, and an arena in which to communicate.

My task in this paper is to address some issues associated with computer graphics that are of particular interest to designers. In this context, I have taken my task to be the articulation of a framework and a set of distinctions which will be useful in a serious consideration of computer graphics and design. It is my intention that this framework and these distinctions will be useful in interpreting and considering more specific issues and particular computer systems.

So that you are appropriately warned, let me say that most of my considerations will be based on a people-centered perspective of computer graphics and design. There are a number of reasons for this. For one, I am trained as a cognitive psychologist and my research has centered on the visual communication aspects of a number of different media. I have, for example, extensively studied the effectiveness of architectural media in explaining proposed designs. I have also investigated a number of computer graphic and videodisc-based systems used in the geographical domain to explain particular places. And, I have considered how computer graphics might be used to emphasize the visual characteristics of mathematics and to enhance the explanation of this complex conceptual domain.

A second reason that I take a person-centered view is because I think this perspective must be emphasized if the complex systems we are now designing are to succeed. It is clearly not enough to design graphic computer systems that can produce attractive pictures; one must also seriously consider how individuals interact with these machines to produce pictures that are suited to particular tasks.

A third reason to take a person-centered perspective is that though this perspective is often mentioned, it is not typically taken too seriously. Frequently, for example, a "user friendly interface" is something that is attached to a system at the last minute to enhance the product's marketability. It is my judgment that the consideration of human interfaces must be made early in the computer system design, and that it is a topic that requires a good deal of attention and care.

A fourth reason to take a human approach to the use of computer graphics is that the advances in developing graphics machines have been so substantial in recent years that the luxury of considering a human view is finally possible. There are now computer graphic systems available, for reasonable costs, that are reasonably manipulable and that enable the production of good quality images. Gone are the days when all interesting manipulations had to be done off-line, as are the days when computer-generated products were clearly inferior to those created by craftspeople. (Figure one.) The challenge is now to find good uses for computers. For, in theory, computers can do anything; they are simply input-output devices. The challenge is to find applications where the relationships between the inputs and outputs are interesting and comprehensible to the practitioners in particular fields.

To begin this person-centered consideration let's examine the design process to see what in the newly developing computer graphic media is relevant to design and designers.

At a basic level, the design process can be described as a set of activities that produce a tangible product. So, for example, the process of sketching and then building a new chair is described as a design process. It is basically characterized by initial thinking, the generation of some form of representation, and then construction. There are a wide range of products that come out of different design processes, including chairs, tables, packages for soap, books, gears for machines, buildings, posters, airplanes, advertisements, toys, and museum exhibits, to name a random few.

Figure one.

Early computer systems were extremely cumbersome. The distance between the systems and the designer was huge. Systems were typically time-share, input was done by batch processing, turn-around time was long, and output was only rarely graphic.

Systems are currently available which make the use of computers in design viable. Systems can be dedicated to a single user, input can be accomplished readily with devices such as keyboards, light pens, and tablets; responsiveness for simple manipulations is rapid and the generation of attractive and detailed graphics is possible.

Future systems, currently available only in research environments, will provide even more accessible and manipulable displays for the designer. Computer workstations can be much like personalized drafting tables except that they can provide for rapid turn-around of images and a single location for a range of activities. The challenge is to provide design specifications for such workstations that are directed toward the enhancement of designer's thoughts.

One way that computer graphics can be used directly is in generating a final product. In current times this product bears the "computer look" which is typically rather sleek and futuristic-looking. As the medium becomes more sophisticated, its products will not have such an identifiable, unitary image; instead, computer graphic systems will become general production tools, much like cameras or pencils, capable of making diverse products and images. In this class of computer-aided design products, we now have some advertisements and movies in which computer graphics have been used extensively to help generate innovative pictorial displays, both still and moving. To a lesser extent, we also have computer graphics used in dynamic and interactive contexts, in some museum exhibits and vending machines for example. In these contexts these graphic systems are not simply production tools, but are instead used by designers to provide audiences with responsive graphic dislays. And, of course, there are video games, a nontraditional design domain which provides audiences with highly interactive graphic experiences.

A second way computer graphics can be used, somewhat less obviously, is in the process of design. Computer graphics can provide tools for designers to use in the development of all products, whether or not these products are generated directly from the graphic system. Computer-aided-design (CAD) and computer-aided-manufacturing (CAM) systems have already been specialized, for example, to enhance the abilities of designers to deal effectively with the representations involved in the design process, and ultimately to transfer these representations directly into production plans.

Interestingly, most currently available CAD/CAM systems are used in conducting routine aspects of the design process. Partly this is to justify their cost, in that repetitive activities are inexpensively accomplished by computer whereas they are quite expensive to accomplish with human labor. In addition, it seems to be easier to attract the attention of specialists to computers than designers, who are creative and seem quite unwilling to endure the inconveniences of current day computer systems. This means that most CAD/CAM systems are being used in large projects, and are specialized for the later stages of design when well specified representations are manipulated and manufacturing schemes are set forth.

Yet the initial phases of design, the initial thinking and the early, informal representations, are also amenable to the use of computer graphic systems. In fact this use will probably become prevalent in a range of generative domains as these systems become more widely available and increasingly approachable by the naive user. In these early phases, the speed of computer graphic production and manipulation will make possible processes of thinking that have not been possible without computer engagement. In contrast then to systems which enable one to accomplish something typically done by humans somewhat faster (or which make certain processes automatic as is the intent with systems which take an artificial intelligence approach), these newer systems will attempt to provide assistance to designers by providing pictorial displays which are not currently possible with traditional media. The attempt will be to enable the designer to keep up with his or her own thought processes, and to view the implications in a detailed and highly manipulable visual display in a very rapid sequence.

There are three representations that become relevant to the use of computer graphics in these early design phases. (Figure two.) There is first, the representation on the computer display that may be the sketch of a building plan, the diagram layout of a book page, or the simulation of a museum exhibit. This display representation is, in the computer domain, always changing. It is the snapshot of an image idea becoming increasingly more coherent as it is worked over time. It is the moment on the stage where the interactions between the designer and the machine are performed. It is the pause, or the retort, in a pictorial conversation the designer is having with himself about a particular design problem. It is the moment in the designer's attempt to explain to a client or colleague what is being accomplished on a problem. Unlike the traditional media, where attractive and detailed representations are too expensive in terms of time and energy to be taken casually, the rapidly changing nature of computer displays allows representations to be changed facilely to match new insights.

A second representation is the machine representation, which is the state at any moment of a system's view of a particular problem. This representational domain is most appropriately the domain of the computer scientist designing a system. Yet it is also relevant to the designer. Depending upon the particular machine representation and the way it has been designed to handle a particular task, there are different constraints on the

Figure two.

Design with a computer graphics system requires the orchestration of three representational systems. These include the designer's mind (the mental representation), the picture displayed by the computer (the display representation), and the computer code (the machine representation). These are not necessarily identical, but they need to be responsive to one another. The display representation can be thought of as a membrane that enables the changing of the different representations.

manipulations that are convenient or even possible to perform. As an extreme example, if the facade of a house is represented in two dimensions only, it will not be possible to conduct three-dimensional manipulations on the representation. Similarly, if the outer surfaces of a house are represented separately from the surfaces of inner rooms, there may be difficulties in uniting the representation of inner and outer details. More abstractly, the choice of any machine architecture will facilitate certain manipulations and make others more difficult. So the intentions of the designer and the purpose of a design environment need to influence the machine's capability to create representations. And, the designer must be aware of the limitations and capabilities of any particular machine and its construction of a machine representation.

The third representational domain which is important to consider is the mental representation of the designer. (Figure

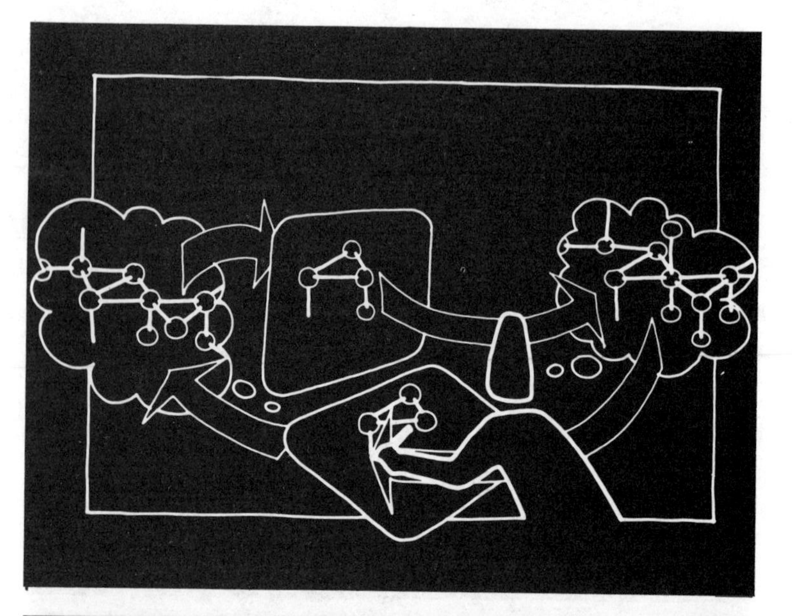

Figure three.

The interaction of the three different representations is central to design. Each idea illustrated on the display alters the machine representation, which in turn changes the display. The new display prompts a change in the mental representation and cues the designer to alter the display in a cyclical process.

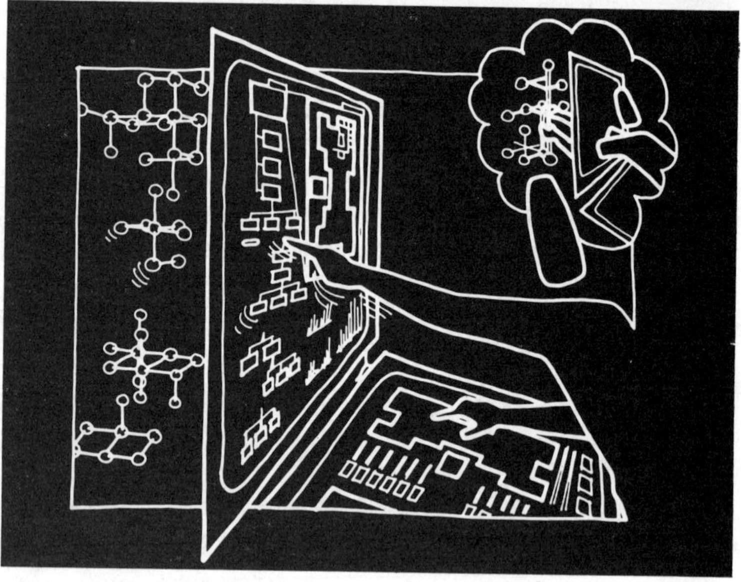

A computer graphics system should provide the designer with easily manipulable parts so that the machine representations may be changed easily to responses the designer wants.

three.) This representation includes the past experiences of the designer as well as the current formulation of a particular design task. Unlike the other two representational systems, this representation need never be made explicit. The better able the designer is to act on the basis of this representation, however, the more effective the design process. In this sense the entire design process is directed by the representation in the designer's mind, as it is articulated first in the display representation and then in a machine representation, and in a responsive new display representation which will influence the mental representation. As alluded to in the title of this talk, the changing pictures made possible with the power of the computer can enhance the process of changing the mental representation (the designer's mind) and hence facilitate progress in design.

The criteria for the effectiveness of any particular computer graphic system rest with a consideration of the mental representation. For the machine representation and the display representations need to be related in a way such that the designer can easily change the machine representation via the display. Elements in the display must be highly manipulable; they must act as programs for action as well as simple "pretty pictures" or elements of a particular view of an object. They must allow the designer to rapidly change machine representations of an idea so as to provide responsive new displays, displays which reflect the changing views of the design problem and which allow the rapid evaluation of its current form.

There are a number of advantages in using a computer graphics system in the early stages of design. Most important, by manipulating and changing the three representations problem solving can be enhanced. The use of the system enables the designer to make an idea explicit. And once it is specified explicitly and systematically, standard manipulations (e.g.

changes of viewpoint) can be accomplished quickly. These rapid transformations in the design enable the designer to encounter specific difficulties with a particular design solution early and make the appropriate changes. They enable the designer to consider a range of alternatives which would be too cumbersome in a non-computer-assisted context. In addition, use of a graphics system frees the designer's mental capabilities; as each piece of the design is specified, full attention can be directed to other aspects of the design without the need to mentally track earlier solutions. And, any level of detail can be considered; studies can be directed to analyses of detail at single levels or between levels.

The implications of the use of computer graphics in design are not, however, as straightforward as this discussion implies. There are subtleties. The most obvious way to think about the use of computer graphics systems, for example, is in the context of self-dialogues, where the designer employs the spatial organization of the display to carry on a conversation with himself, or herself. Yet these systems can also be used to communicate with others, be they clients or colleagues, be it in the same office or across the country via telecommunications systems, be it a formal presentation or an exchange of ideas in an interactive context. Though it is simplest to imagine the solitary surround of such a system, it is also important to realize that it sits in a social context, much as a drafting table sits in a design studio. For the development of graphic routines and expertise on the system is inherently a social act, as a number of individuals contribute and share their expertise.

Similarly, it is easiest to imagine the designer sketching at the graphics workstation, much as is done at a drafting table. Yet it is important to acknowledge that graphics systems also enable the designer to select images from a set which has already been stored in the system (or which are available on a peripheral system, for example, a videodisc system). These stored images can then provide input to the design, and be available directly for modification. In the same way, textual sources of information can be made available on the system so the designer can also research those aspects of a project that are not imagery-based.

Because the designer can use the system to store thoughts and sketches on a project at any point in the design process, it becomes highly useful for documentation. These stored thoughts can be retrieved later in the process to make explicit the reasons for a particular decision, or to review the progress of problem solving. Such a capability also enables the designer to frequently put aside parts of a project to be referred to later without being concerned about forgetting the relationships between a particular set of ideas and their bearing on a problem.

Computer graphic systems can then provide designers with access to a productive virtual reality in which to manipulate their initial concepts in the design proccess. These systems can act as assistants, providing the designer with extensions of his or her own abilities, and access to a range of materials relevant to design. They can effectively enhance the changing of pictures in the early stages of design to change the designer's mind.

These considerable powers are in addition to the graphic system's capability for helping to produce formal representations in the later design phases that make the early, loose images more articulate.

The use of computer graphic systems as production tools and as representation manipulators are somewhat obvious uses of these systems, and are current realities in some cases. A third perspective on computer graphics and design is less obvious and considers just what the tradition of design has to offer computer graphics and, more specifically, what kinds of opportunities there are for designers in the design and development of graphic systems.

There are basically three classes of experiences in design that are relevant to computer graphics system development; the design of environments, the design of activities, and the design of catalysts for attitude change. The design of buildings, for example, requires that the designer produce an environment in which a wide range of actions will occur and a large number of individuals will interact. This experience is extremely relevant to the design of computer workstations, places which will also be intimately used by a range of different individuals for a range of purposes. Social, functional, and individualized issues need to be addressed in this context, much as they are in the design of a building.

A second relevant experience is in the design of activities, of toys and playgrounds, for example. This design experience provides designers with a sense of how to encourage activities, how to design a situation that invites involvement and maintains attention. Such a perspective is central to the design of graphic systems.

A third class of relevant experience is in the design of information graphics. For computer graphic systems make possible interactive books, interactive movies, interactive exhibits, and so on. The design of these media will require the integration of text and graphics in a temporal domain. Experience with issues related to communicating information will be clearly relevant to such new classes of design. With computer graphics, new classes of design tasks will be provided, as well as assistance in the more traditional design activities.

A simple definition of computer graphics is "a set of images generated by a computer." Though basically accurate, such a definition provides only a very shallow basis to understand the implications of computer graphics systems for designers, and only a spare framework for evaluating the different systems. In place of this first level definition, I have attempted in this paper to substitute the notion that a computer graphic system is "a machine-based system which provides a setting for graphic interactions," interactions which can enhance both the early and later phases of design. Through interactions with people and through changes of displays, machines can mirror changes in human minds and make clear the implications of such changes. The design of settings which can encourage these interactions is in itself a design task, a fact which makes the whole process interesting, exciting, and worth engaging in.

The Know Business Is Show Business: Graphic Design and Computer Design

Aaron Marcus

Computers have three components, or "faces", and designers have important roles to play in the structure of each of them.

In the past two years I have focused my attention primarily on the intersection of computer graphics with graphic design. I see three important relationships between traditional graphic design expertise, especially in information-oriented graphic design, and the needs of the new computer graphics technology. Now that computer graphics has proved that it works, we can ask ourselves; What do we want to show. What is significant How can we show it well. How can we show it effectively? The name of the game in the Image of Information Age is this: Know Business is Show Business. Graphic designers clearly have important contributions to make. There are three specific intersections, or in computer jargon, interfaces: Outer-Faces, Inter-Faces, and Inner-Faces.

Outer-Faces

Outer-Faces are the images of information that computer systems produce: texts, tables, forms, charts, maps, photos, and diagrams. In many cases these images are poorly designed to meet the communicative goals for which they were created.

Graphic designers can help make more effective images of information. Even more imporant, they can help to transfer some of their expertise directly into the software so that the computer system itself can make at least reasonably good designs.

The computer graphics systems can in fact produce such images. Graphic designers, working in cooperation with the producers of such systems, can help the systems to do their job better. As Karl Gerstner proclaimed 15 years ago, graphic designers will design processes, not just products.

This will not put graphic designers out of work. The needs for responsible information graphics of all kinds are so great, and so under designed in our present society, that there is enough work ahead for both human and machine to keep everyone challenged and active.

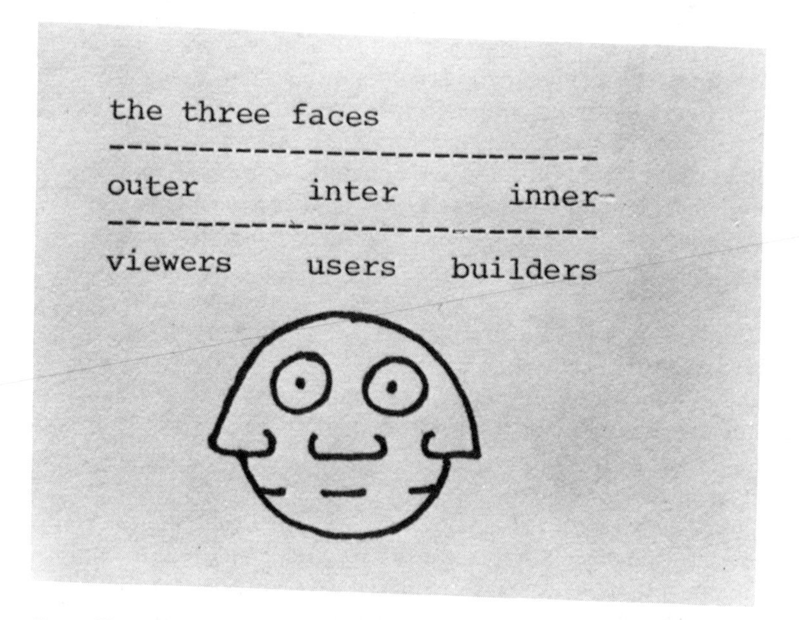

Figure One

The Three Faces
The three faces among design and computer graphics are the Outer-Face, the Inter-Face, and the Inner-Face. The Outer-Face is the image of information that computers produce: charts, maps, texts, and photographs. The second face is the Inter-Face between human and computer, or the means by which the user communicates with the computer. Inter-Face devices include: keyboards used for input into the computer, touch sensitive CRT displays, light pens and tablets. The third face is the Inner-Face, or those elements that make the intricacies of the computer understandable to the people who build and maintain it.

Inter-Faces
The second face is the human-computer Inter-Face to the machine. By this I mean the frames of information the computer presents to its user. These may be verbal questions and answers, or they may be diagrammatic visual structures the user can touch on the screen in order to indicate what decisions are being made. The interface often includes lists of contexts of data stored in the computer so that the user can see what should be done in order to create a final image of information (an Outer-Face).

Here, the design task is an enormous one. In the past, printed books presented information to decision makers in a linear fashion with one point following logically out of another. With computer systems, the dynamics are far more complex. The computer stores frames, or pages of information, each connected to the other in various ways and on various levels, forming an information network. You do not have to read this electronic book in a serial fashion, but can jump from point to point. In an odd way, we can see that once books were made from trees, but now books are becoming "trees" with branches and roots of information. Designing the frames of information that make up the tree will be the second major task for graphic designers. We will see widespread use of such systems soon in the new teletext and videotex (or viewdata) systems which are popping up world-wide.

With the job of designing frames of information will come the task of designing training material to teach new users about these computer systems. In the future, we will all be computer users, and the problem of developing computer literacy will be a huge one.

Inner-Faces

The third face is the Inner-Face of program visualization; all those means by which a computer system makes itself known to the people who build and maintain it.

Consider programming languages: most of them are written to be printed on a typewriter or line-printer device and ignore all of the typographic richness available to typesetting equipment. A third role for graphic designers is to help determine how all of these programs can be visually enhanced, or to design entirely new means of indicating the structure and process of these complex systems. There are no well established rules for doing this. It is a new and uncharted territory, for which we'll need new kinds of visual-verbal diagrams.

With respect to program visualization for the maintenance of computer systems, this is no small project. Much of the expense of computer systems is now shifting to maintenance, and this will be an increasingly important aspect in the future. It might sound somewhat technical and dry, but the wildest dreams of typographic and symbolic form might in fact become useful tools for taming and maintaining complex information systems.

Applications

I have had an opportunity, in working with state of the art computer graphics systems, to explore all three of these facets of visualizing knowledge. I held a staff position as a research scientist in a major government laboratory, Lawrence Berkeley Laboratory (LBL).

At LBL I helped to design a large data base and information management system called Seedis. It is housed in mini-computers and exists as a network across several locations in the U.S. Information from the 1970 and the 1980 censuses will be contained in it as well as other giant data bases. The information can appear as charts, maps, reports, etc., semi-automatically constructed by the data processing and graphics modules of Seedis.

Opportunities to contribute graphic design expertise have arisen in several ways. I advised the computer science and mathematics department which originally built Seedis on how to make better, more effective charts and maps. Some of these suggestions have been translated into the software so that better images of information emerge as the default capabilities of the system.

Another major activity has been to help re-design the interface for Seedis and its on-line help facilities. This means re-designing how frames of information look when a user is interacting with the computer. Such matters as capitalization, gridded layout of text passages, style of punctuation, visual and verbal consistency are important in order to provide a so-called "friendly", considerate, well-designed interface to the system. I have also been active in helping to write some of the explanatory texts which can be called up at any time by the user

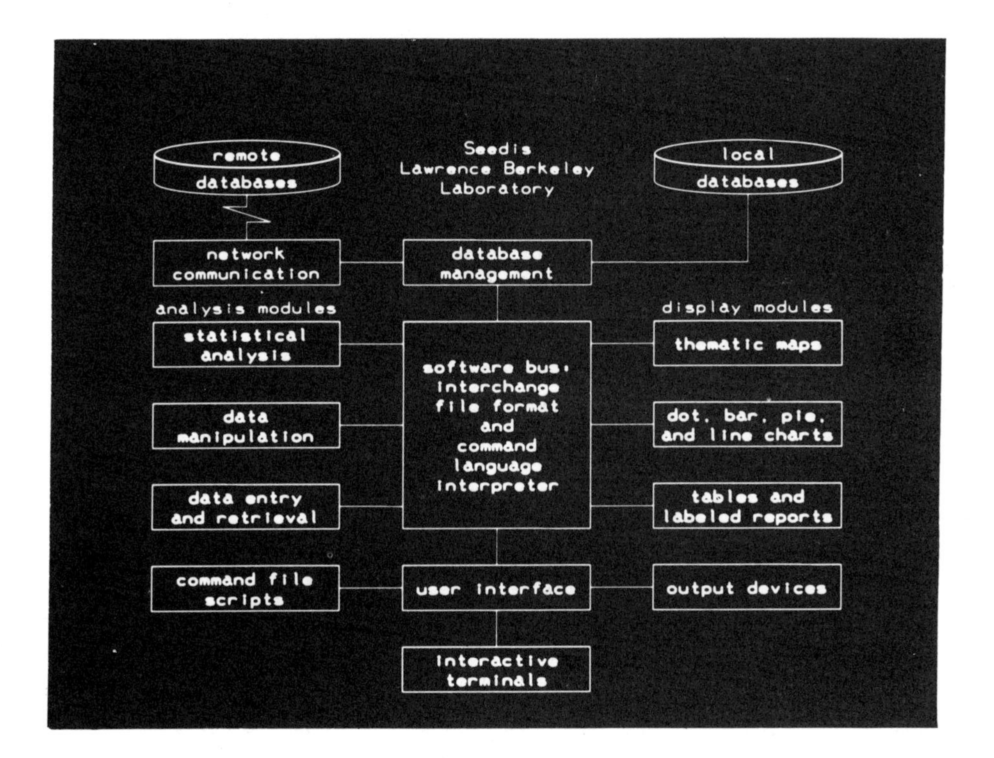

Figure Two

The Seedis Project

In the Seedis project, the U.S. census data from both 1970 and 1980 will be housed in a large data base. Links will connect the computer system to stations across the country. The computer will conceivably be able to generate books of information in response to direct requests.

The charts, maps and diagrams generated by the computer will be in part, created by the computer itself. Directed by basic design programs that dictate clear and concise information visualization, the computer will be able to "create" texts and diagrams. In this way, the computer system can be responsive to specific input, generating those things specifically asked for by the user.

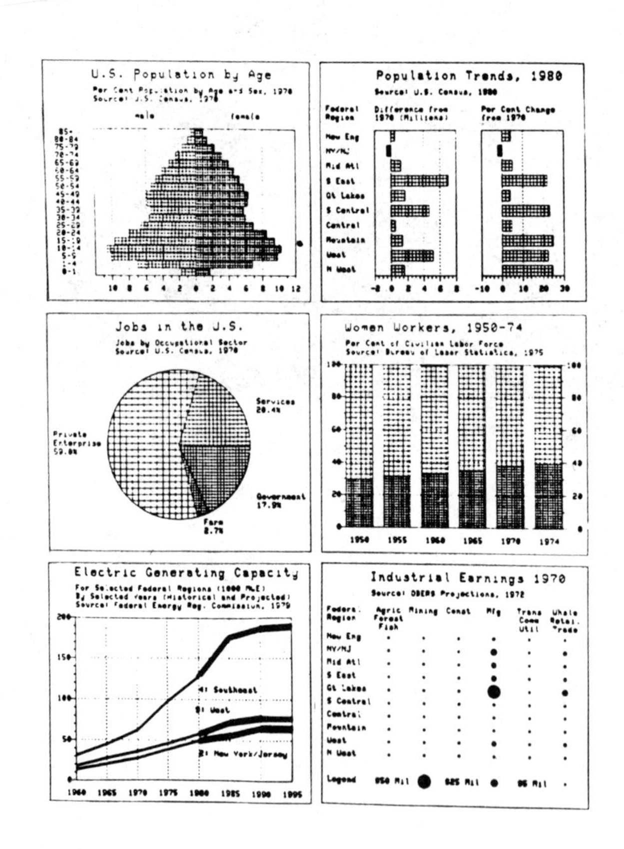

and displayed on the screen.

One of my projects is especially interesting. I have been drafting a graphic design manual for Seedis. This information management system will contain the specifications for symbols, spatial layout and temporal sequencing that Seedis utilizes. It will also contain general information about graphic design. That information will be useful to computer-based people who lack graphic design experience, but who have an interest in the subject and are actually making graphic design decisions about typography, color, and the like all the time.

Two-million-page book

A related project concerns the production of a "book" with 2 million pages. Well, not actually a single book. Using the capabilities of Seedis and the data from the 1980 Census, the Seedis staff are active in a project that will allow individually assembled groups of charts, maps, texts, and tables to be produced for thousands of different governmental offices in the country. These will be produced on a laser typesetting and printing device and distributed by mail.

In effect, anyone can have a book made to order. The complex pages will be automatically laid out and typeset as complete images by the Seedis information management system. The project calls for designing complex standards for charts, maps, and tables.

I have also initiated research projects in program visualization which will keep me busy for several years. This kind of project, more than others, concerns research into the future of visible languages.

The scale and complexity of these projects are staggering. The implications they have on society are awesome. I have been trying to alert the graphic design community since 1968 to the

I am actually doing those things I might only have dreamed of years ago. Sadly, at the moment there are only a handful of graphic designers who have made the switch to the 21st century. Luckily, with the increasing availability of computers, more and more people are becoming involved with computer graphics.

Designers progress slowly
Even though home computer use is growing, I am concerned about the slow progress that art and design schools and smaller design studios have made in trying to work with new media, new technology, and the new social context for graphic design. At the University of California at Berkeley I created a course called "Maps and Diagrams" (1980) in which all of the students used Apple home computers and television sets to do their graphic design work. Slowly, other schools are developing the expertise and curricula.

In the meantime, I see projects that go begging for graphic design input, but there are not enough graphic design people knowledgeable in computer graphics to meet the needs.

I take two views toward the solution of the problem. First, graphic designers are going to have to become more involved with computer graphics. Second, computer technology people will have to become informed about graphic design. Surprisingly, I have found more enthusiasm for graphic design in computer science than vice versa.

As far as the future is concerned, the sky is the limit. Computers communicate with people primarily through visual symbols. It is just common sense that the computer world should work with the graphic design world to help determine and refine the nature of that visual symbol communication.

This is the essence of the crucial role which graphic design can play in emerging human-computer technology. The goal, as always, is a humane mixture of art and science, of reason and intuition, of word and image, of information and significance.

The Interface to Design

Eric A. Hulteen

Computers may be smart, but they're not very intelligent. The users of computers are forced to learn the language of computers in order to harness their power. A "human interface" would force the computer to understand the way people communicate.

Computer-aided design systems (CAD) can potentially serve as a powerful tool for the designer. As it is however, few designers have the inclination or the education to make use of CAD. This is not a failing in designers so much as it is a failing in the computers. What is generally required to use a computer is a language that is foreign and unnatural to people.

Ignoring the problem will not help. Designers need to learn what kinds of man-machine communication channels are available today, and will be available in the future. Advances in computer-aided design will be made more quickly if designers know what they want from the systems and can tell the manufacturer. The form those advances can and should take are those allowing a communication that is both natural and comfortable to take place between man and computer.

Right now the principal uses of information processing systems in design firms are, for the most part, number-crunching operations necessary in the management of any business. Some computers can and are used in what could be called "technical" design: engineering calculations, graphic manipulations, and drafting. But, for graphic design and architectural design, computers have yet to make a significant contribution.

The designers' plight is that while computers could help them in many ways, the difficulty in communicating with computers stands as a barrier to their use. The answer to this problem is an interface between designer and computer that is easy for people to use, a human interface.

Some say this interface problem is being addressed, but the results so far have been - to say the least - inadequate. The manufacturers of CAD equipment generally claim that their systems are "highly interactive", yet the list of devices people may use to communicate with computers seldom includes anything more than light pens, tablets and keyboards. All of these are one step removed from normal human-to-human interface techniques. A keyboard requires the skill of touch typing, tablets and light pens are certainly more difficult and time consuming than simply pointing at an object. The manufacturers are quite correct in thinking these design systems need to be highly interactive, but they have not yet succeeded in making them that way.

Computers - not humans - need to learn another language. Computers need to understand the more subtle nuances of human communication, paralleling as closely as possible, the style of communication between people. Instead of designers learning to communicate with a computer on its terms, the computer must learn to communicate with people in human terms.

It is easy to understand the importance of human channels of communication and the value of equipping a computer with those channels, when one considers what it would be like to be without them. Imagine a day at the office when it is useless to speak because no one is able to understand speech; when there is no reason to point at something to which you are referring because no one can perceive that your arm has moved, or where you are pointing; a day when no one talks to you because they don't know you can hear; or a day when no one is sure whether you are paying attention because they don't know where you are looking. The absence of these natural communication channels would likely destroy the possibility of accomplishing any creative work. Consider the members of a design team trying to work when the principal way to communicate with each other is through a keyboard. The future of computer-aided design certainly has more to offer than that.

There are devices available - although they are under-used - allowing people to communicate with computers in more familiar ways. The most important of these input channels is

Figure One

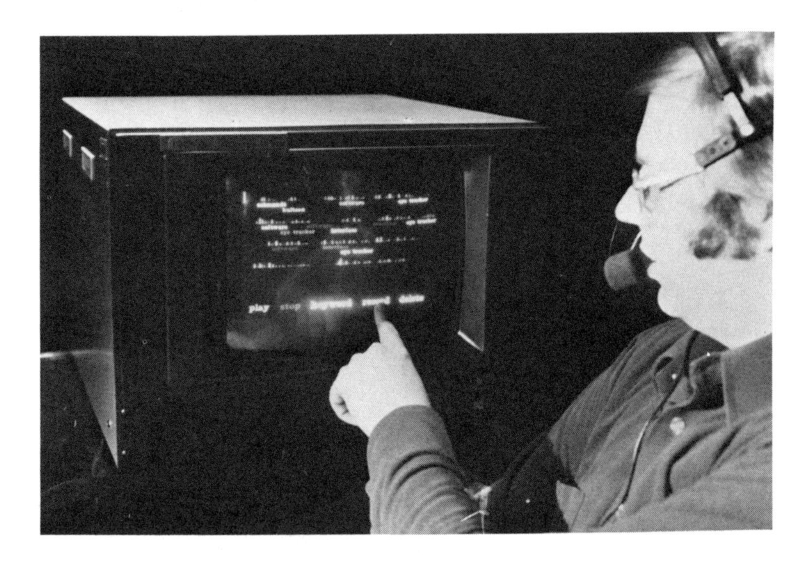

Speech Recognition and Touch Sensitive Display
The user is speaking into a head-mounted microphone connected to the computer's speech recognition system. The computer can match the spoken word to those words it has stored in its memory to "understand" what it is being instructed to do. At the same time, the user is touching a monitor with a touch sensitive surface for direct interaction with the image.

speech. For the computer, speech recognition is essentially a pattern-matching activity. The systems record the way a word is spoken during the training process, and save that as a reference. Then, in the recognition mode, the systems compare what was actually spoken to the stored reference. If they match closely enough, the word is considered recognized. None of the systems are 100 per cent accurate, nor are they ever likely to be: there will always be difficulties in matching the wide variations in pitch and intonation. Despite imperfections, the use of speech recognition would mean an enormous improvement in the relationship between human and computer, allowing people to use one of the most elemental means of communication.

Another device would permit people to use touch to communicate with the computer. To make a CRT screen a touch sensitive display (TSD), a means must be provided for the computer to determine if and where the display is being

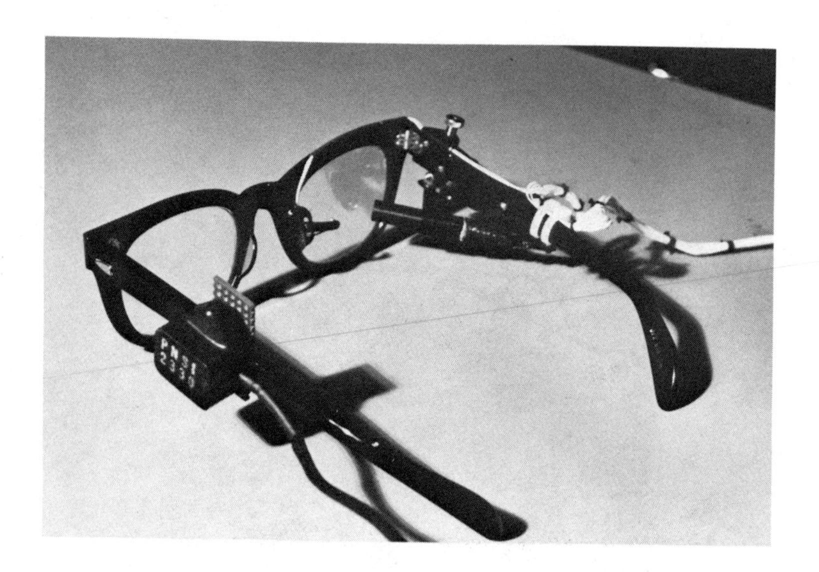

Figure Two

Eye-Tracking

This eyeglass frame has eye-tracking equipment mounted on the right arm, and a three-dimensional digitizing cube mounted on the left arm. The eye-tracker reports the attitude of the eye with respect to the head, and the digitizing cube reports the position and attitude of the head with respect to the room, allowing the computer to trace the point of regard.

Figure Three

Gesture Recognition and Speech Recognition

This user wears a three-dimensional digitizing cube strapped to his wrist, allowing the computer to track the arm's location, and follow where he is pointing. To verify and reinforce instructions, the user also wears a head-mounted microphone connected to the computer's speech recognition system, so he may talk to the machine.

touched. The TSD is simply another more direct form of communication that eliminates intermediary instruments like light pens, tablets and keyboards. It is unnecessary to pick up a light pen when a finger will do; nor is it necessary to coordinate one's movements on a tablet with their effects on a screen if input can occur directly on the screen.

The TSDs and tablets are two-dimensional digitizers. They convert the position of the light pen on the tablet, or the finger on the screen to X, Y coordinates for the computer. One step beyond these are three-dime sensor and "see" where the person is pointing. Pointing is a gesture both instinctive and essential to human communication. The fact that a computer can be taught to understand it, further narrows the gap between human and machine.

The same is true when a computer can determine where one is looking. People look at objects to which they are referring in the same way they point or gesture at things. Computers should be able, just as humans are able, to read and interpret a look. Eye-tracking technology is relatively new, but there are several types of trackers available. The more inexpensive and less accurate of them are now being used primarily to help handicapped people communicate more effectively. They are usually mounted on an eyeglass frame. The more expensive ones, in addition to being quite accurate, do not encumber the subject at all. These trackers are TV cameras which sit on the floor several feet in front of the user. Most eye-trackers work by bouncing infrared light - which we cannot see - off some part of the eye and by monitoring the reflection.

Just as important as the computer's ability to understand us, is our ability to understand the computer. For this, the machine must be able to output information to people in an intelligible way. Speech is the most easily understood means of communicating for most people, and it is becoming simpler and less expensive to give computers the ability to speak. There are a large number of speech synthesizers on the market, and their prices are falling dramatically even as their quality improves.

Given the availability of these and other input-output technologies, the question becomes how the human interface is to be implemented. There are four characteristics that such an interface requires. First, it needs to be interactive. It must be possible for a conversation to take place between the designer and the machine. Communications must work easily in both directions with both parties making contributions to the design process and talking things over when something is unclear. An essential requirement for interaction is that users have a clear idea of what the machine has already understood. This means that feedback needs to be provided indicating to the user that the machine really knows where he is pointing, what he is touching, or where he is looking.

Second, the interface needs to be intelligent. The computer needs to bring all the information at its disposal to bear on any issue in an effort to solve it alone before it bothers the designer with a question. Some of the needed information might come from alternate input channels. If, for example, the computer receives a spoken input that has information missing, it should not ask for the input to be repeated. It should determine what information is missing, and try to get the information in some other way. It could check where the user is pointing, or looking, and only if those fail, ask a very specific question which implies how much it already understands.

The system's "intelligence" can also grow out of the application of some rules about what allowable forms inputs might take. By doing syntactic and semantic analysis on spoken input, for example, the computer is able to generate intelligent feedback for the user. If the input received says to create a square, the computer must observe that no color was specified. Then an intelligent question might be to ask, "What color?" Intelligence can also come about by the application of contextual information. The interface software could, for example, prevent the user from putting two objects in one place, or if it notices a reference to an object that does not exist it might assume that it should create one. Intelligence can be exhibited by the machine not asking unnecessary questions, by making valid assumptions based on context, or by using alternative sources of information when necessary and available. The important thing is that the interface behave as an intelligent co-worker so that a designer can interact with it in a natural way.

In addition, the interface needs to be idiosyncratic. That is to say, it should respond to the designer as an individual, whoever he or she might be. One person may prefer to talk about things in a different way or use a different language than someone else; another may prefer a graphic output to a verbal one; some may want the computer to maintain a lower profile or a higher one; and some may require detailed references. The interface needs to respond as another person would, to the individual characteristics which make each user unique.

Finally, the human interface must be multi-channeled. Such channels should include speech recognition, gesture recognition, eye-tracking, speech synthesis, TSDs, tablets, keyboards, or any means that people can employ to aid them in communicating. There should be as many ways as possible for people to communicate with the machine. One assumption behind all work on the interface is that no single channel will ever be 100 per cent accurate and that it will always be necessary to clarify inputs by using data from other channels. When people communicate with other people, it is very common for them to use more than one channel simultaneously presenting redundant outputs. An example would be when one points at a salt shaker, looks someone in the eye and says, "Please pass me that." If a computer is prepared to understand all of these gestures it becomes much more likely that its interpretation of the input will be correct. In short, redundancy aids accuracy.

Having alternative input channels also makes the interface easier to use. Given a situation where one input channel is already saturated, the existence of an alternative channel makes it possible to bypass the congestion and continue operation. An example would be that of a CAD system operator using a keyboard as an alternative input channel because he is already talking to someone.

The more general point to be made here is that design should be viewed only as a metaphor for any sophisticated, creative work to be accomplished with the aid of computers. The need for a human interface by no means applies only to the field of design. As more and more powerful information processing systems become available, what the machines can do becomes not so important as how easy it is to get them to do it. Recent increases in the functionality of computers have not been matched by increases in their useability. This trend must be reversed or some of the enormous potential of computers will simply be wasted because many people simply won't make the effort required to use them.

Coping with Complexity

Charles A. Mauro

Information demand and user advances in hardware and software are making it technically possible to process more information. However, the human operator's ability to process information is remaining fixed. The result—user frustration and dissatisfaction.

Ours is an age of astonishing technical achievement. The products we create are capable of more than ever before. But, realistically, we are not reaping sufficient reward from our prowess. We trip over our own achievements, our technology.

Computer technology dominates our times. More and more, products like communications systems, control and recording systems, data generating and "crunching" systems are based on computer technology. Yet, too often, these sophisticated products and systems are poorly designed for their users. Non-usability of design extends from commonplace consumer products to large systems like control rooms of power generation stations.

Usability problems and their solutions are as new and complex as the technology. Solutions require entry into fields of knowledge unfamiliar to many designers. The problems are unremitting; and they will continue to grow worse unless designers accept the challenge of designing for the dimensions of the human mentality in the same way we design for the dimensions of the human body.

To understand the sort of problem faced in designing information systems we must eliminate the widespread fallacy of an "Information Explosion." (Computers are now more numerous than the human beings of our planet.) Computers enable us to process more data, more rapidly than ever before: we have created a "Data Explosion."

The term "Information Explosion" is a misnomer; data and information are not the same. Human mentality discriminates between data and information; the human mind discards nearly all the data that is available in order to extract relevant information. It is estimated that in the usual course of events we discard 99.998 per cent of all the data which we encounter. If we didn't, we would be extremely ineffective in making any choices or decisions. Data is silent, like the unobserved tree falling in a forest, until attended to by the mind. It then becomes information.

For example, the face of a conventional watch contains data in the form of numbers from one to twelve. Yet, to find the time of day, we need only consult the information which corresponds to the position of the two hands. Telling time on a digital watch involves a similar process. Faces of some digital watches are more data-packed than conventional watches. To get a quick reading we discard all visual elements of the face except the digits which indicate time.

The distinction between data and information is fundamental when designing systems for human use. Possibly the most important question of our day, next to survival, is how to develop and design systems which facilitate the transfer of data into information. A product's effectiveness is heavily dependent on meeting this criterion.

Few products measure up. Computer displays, for example, are seldom presented in a clear format with data selected and organized to optimize the decision making process. The norm is for displays to be cluttered and confusing. Too many designers believe, when it comes to data displays, that more is better. In psychological fact, almost the reverse is true. The more irrelevant data one can prune from a display, the greater the chances are that the observer will see the essential information.

Until designers really come to grips with this concept we can expect displays to be driven by the data-producing capabilities of technology rather than the information needs of the human user. The result will be to continue and exacerbate the problem of over-burdening the user with the task of extracting the signal he needs from the background noise of endless data.

The computer promises us the potential for making more accurate and more effective decisions. In some cases this potential is being realized. But in many cases the decision makers who must transform data into information are already over-burdened by the problems of the business world. Their jobs are not made any easier, nor their decisions more astute, by contending with a technology which is complicating the data they need to make decisions.

In all cases, the critical link between data and information is the human being. Whether it is a product for individual use or a complex corporate computer system, the object should be the same - to facilitate the human decision making process.

As an example of how large the problem can be - and how clumsily it has been handled - we can look at the operator control rooms of nuclear power plants. These systems are complex integrations of human and computerized functions, but they present the same basic data-to-information transfer problems found in all kinds of product design.

The chief problem in such a system is allocating jobs between the computer and the human operator in the most fruitful way possible. This should mean understanding what tasks are best performed by computers and what jobs are best performed by people. Humans are good at solving big problems; they can look at the system from a distance, evaluate, judge, and make decisions. The computer, on the other hand, is very good at solving discrete, repetitive tasks. Most of the computers on the market today do not try to fit tasks to human performance.

There are three conventional organization formats to resolve the allocation problem. One format assumes the human operator will function best if he is given more data. What is forgotten in this scheme is that while the computer's ability to produce data is seemingly infinite, the human ability to process and digest data is stricty limited - the result, an over-loaded operator. Clearly, there will be errors when the operator is over-loaded; he or she will omit necessary functions and forget others. The trouble here is that the system demands this operator to have a profound understanding of the entire system and its intricacies, including perhaps the discrete, repetitive tasks better left to a computer. Managers of such a plant will discover eventually that they have trouble keeping workers, and training the ones they do get.

The reverse problem is over-automation, a situation in which the computer handles almost everything. In this case, the operator does not need to understand the system, and he or she does not care to. Errors occur here too. Because this operator is unfamiliar with the system, he or she responds badly when problems do arise. Morale in such plants is generally poor and workers are relaxed to the point of boredom.

A third solution seen more and more often is what I call "gap-filling." The system is automated, but for those spots where engineering and technology cannot do the job, the human being is introduced. In a typical nuclear power station, it is common to have periods where the computer is in total control. Humans step in only for those jobs the computer cannot do.

This is in fact, the same sort of thing that occurs in product design. Manufacturers use instructions to make up for deficiencies in the product's human performance design. Manufacturers, just like engineers, fill the gaps with the human operator where the system either could not or should not be automated.

The accident and the system design at the Three Mile Island nuclear power plant serve as a graphic illustration of just what poor function allocation can mean. (Figures one to six.) At the time of the accident, the computer was in full control of the

Figure one.

Characteristic Problems in Allocating Functions Between the User and the System.

On the horizontal axis are charted the various states of a nuclear power system. Along the vertical axis may be mapped the functions allocated to the operator and to the system.

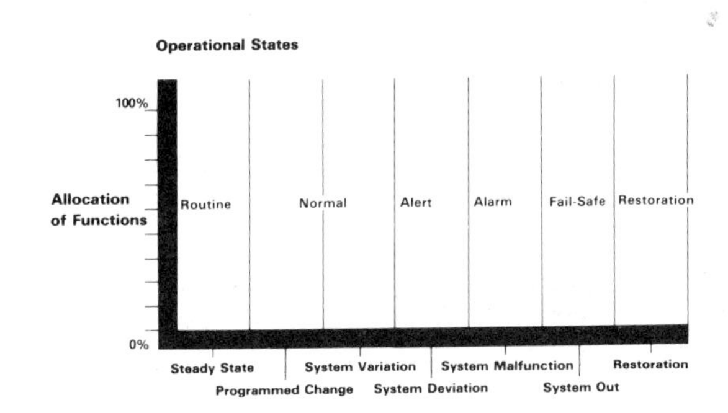

Figure two.

The dotted line in this diagram represents the operator, the solid line represents the system. In this instance, the operator is being asked to perform consistently at an extremely high level. This system requires the operator to have a sophisticated cognitive map of the system's workings. Mistakes and accidents are likely to occur as a result.

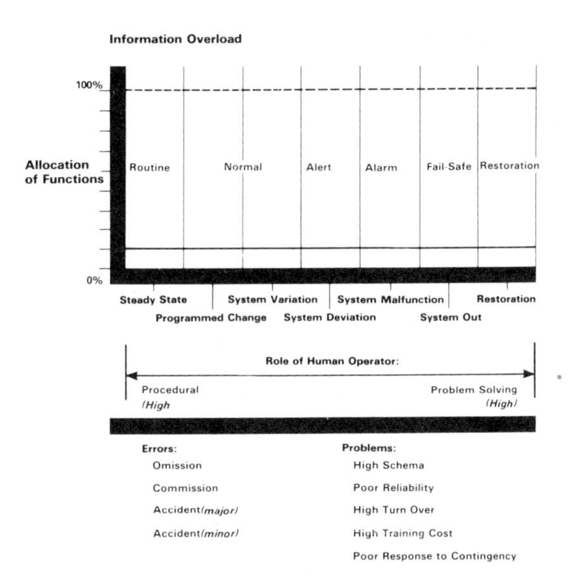

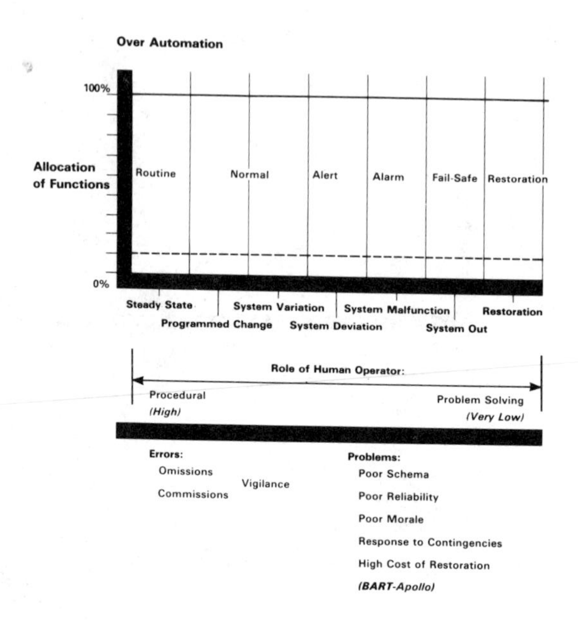

Figure three.

The problem reversed. In this system, the operator has very little responsibility because the system is in virtual control. The operator needs to know almost nothing about the system. Different problems arise from situations where the operator is essentially taken out of the loop: operators have poor morale, poor reliability, and respond badly to emergencies because they are unfamiliar with the overall system.

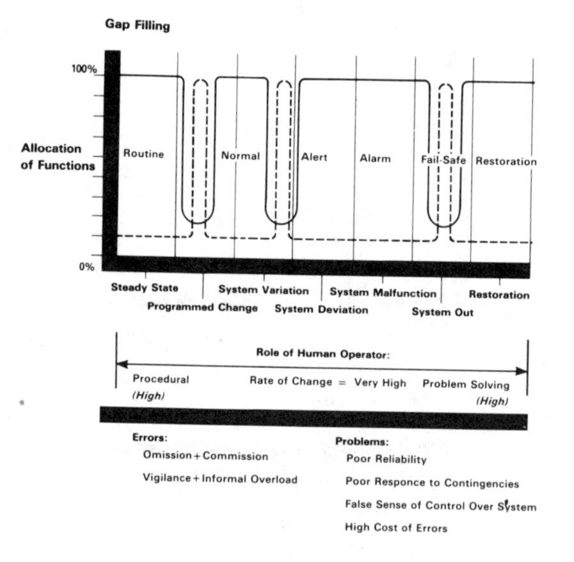

Figure four.

This sort of gap-filling system is becoming a common, if uninformed, answer to the problem of function allocation. It is used when software and hardware designers are trying to build an automated system. In those places where the system cannot be automated, the human operator is thrust to solve the problem. The results are periods when the system is in total control and periods when the operator is in total control.

Figure five.

A diagram of the allocation of functions at the Three Mile Island nuclear power plant before and during the historic accident that took place there. Before the accident, the system was in complete control. At that time, the operators were relaxed and unprepared for contingencies. As the situation became critical, the system suddenly handed responsibility to the operators.

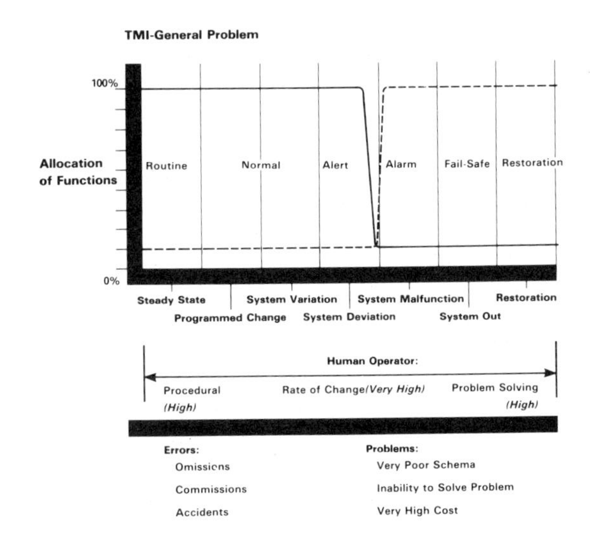

Figure six.

An ideal man-machine system that selectively feeds the operator into the system as the problems become more complex. This sort of system allows human and computer to be used to their best advantage: the human is best at solving big problems; while the computer is skilled in solving discrete repetitive tasks. Allocating tasks to the system and to its operators is an extremely complex problem.

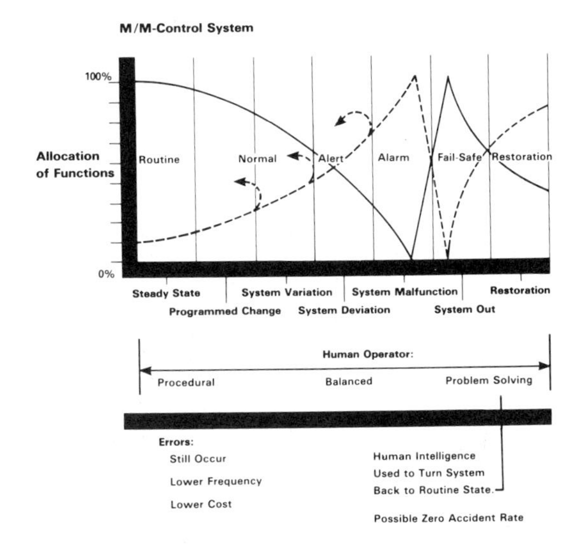

system, and the operators were relaxing. As it was designed to do, the computer handed the system over to the operators as it went first into alert, and then a fail safe mode. The operators were not, at that point, fully prepared to make the decisions required of them: the transition in responsibility was simply too sudden. They were given too much data, too rapidly to make reasonable decisions. The fault was not theirs; it was the fault of a poorly designed system.

Ideally, in the man-machine system, the human operator should be progressively factored into the decision making process as problems become more critical. In this way, the operator would have the time to adjust to the decision making role, evaluate the problem, its seriousness, and act. That is putting human intelligence -and, for that matter, mechnical intelligence - to its best use.

Products far simpler and less dangerous than nuclear plants can and should be evaluated for the same human performance attributes. In the past, both designers and manufacturers could afford to be a bit cavalier about their products - as long as those products sold. But research shows that those carefree days are fast drawing to a close. Consumers now are more sophisticated and less inclined to put their trust in a product or a manufacturer. Consumer trends indicate that belief in advertising is decreasing, while peer group recommendation is on the rise. Now, the post-sale performance of a product may well determine whether or not it continues to sell. Consumers will still appreciate it if it looks nice, but more important, it better work when they get it home.

Over the past five years, my associates and I have developed three measures to evaluate a product's user performance: the user-merit index; judgements; and opinions. (Figure seven.) Using these three to evaluate either a product or a prototype, one can understand how the human operator should be

<u>Measures</u>

-Performance	- User Merit Index
-Judgement	- Semantic Differential Scale
-Opinion	- Structured Interview (and free comment)

Figure seven.

Three measures used by the author to evaluate products' effectiveness in terms of their human engineering.

User Merit Index

- Rates Degree of Difficulty for User
- Assesses Demand for Specific Learning

Figure eight.

The user-merit index (UMI) is based on a simple psychological theory of equity saying one should be able to retrieve satisfaction from a product in equal proportion to the amount of effort put into making the product work.

factored into the system.

Here I would like to look closely at one of the three measures —the user-merit index (UMI). The user-merit is based on a simple psychological theory of equity saying you should be able to retrieve satisfaction from a product in equal proportion to the amount of effort put into making the product work. We factor that theory of equity using three UMI criteria. The class-1 criteria evaluates whether the product fulfills the function for which it was designed. Can you tell time on the clock, or write a letter on the word processor? If the answer to that question is no, the product has failed at its most basic level. (Figure eight.)

For the class-2 criteria we assess, through laboratory observation and controlled experiments, the level of learning required to make the product work, and how difficult it is to make it operate. (Figure nine.)

In the third, and least important of the three criteria, we examine the physical attributes of the product: its color, shape, weight, and so forth.

For each class, the product is given a value, which is plugged into an equation and normalized on a scale of zero to 100. For each of the values between zero and 100, there is a given level of experience for the user, with 100 representing the most basic, and zero, the most advanced. Using this index, it is possible to determine whether the product is suitable for consumers, dedicated users, or very experienced users. (Figure ten.)

What is suprising about all of this is discovering how very difficult it is to make many products work. (Figures 11 and 12.)

We evaluated a digital watch for a manufacturer. The study group was broad, with 56 participants ranging in education from ninth grade to PhD's in electrical engineering. The task was seemingly very simple: to set the watch ahead one hour. Not one person could do it without the instructions. When we gave them the directions, about 43 per cent of the participants managed to set the time correctly. We took the instructions away from that 43 per cent and asked them to repeat the task. Only ten per cent could repeat it. (Figure 13.)

Clearly, this design does not work very well. The watch passed the first UMI test, it is easy enough to read the time on it. But it failed the second class miserably. The level of effort required to set the time was well beyond the experience and patience of the most sophisticated users.

The digital watch is not a total aberration. There are other products on the market now with similar problems, among the most difficult to work are the computer-based products. These products simply are not designed with the consumer's needs or

Criteria

-Class I - Fulfill Purpose
-Class II - Ease and Accuracy of Use
-Class III - Physical Attributes

Figure nine.

The three UMI criteria in ranked order of sigificance.

$$\text{UMI} = 100 - \frac{\sum^{NT} (\text{scores})}{\text{Max. score}} \times 100$$

Where N = Number of Persons
T = Number of Tasks
Max. Score = 3 × N × T

<u>Scores</u> When tasks performed correctly with :

0 - Device only
1 - Short instructions
2 - Long instructions
3 - Verbal explanations

Figure ten.

For each criterion, the product is given a value which is then plugged into the UMI equation and normalized on a scale of zero to 100.

Figures 11 and 12.

For each of the values between zero and 100 there is a given level of experience for the user, allowing one to determine whether the product is suitable for consumers, dedicated users, or very experienced users.

UMI Interpretation

100 **All persons, all tasks, need only device; high self-evidency; no learning, assured reliability**

90 **30% of tasks and short instructions (prompts, etc.,) or 10% of tasks need verbal explanations**

80 **20% short instructions and 20% long instruction or 20% verbal explanations**

50 **From 50% to 70% of tasks need some instruction**

30 **From 70% to 100% of tasks need some instruction**

10 **From 90% to 100% of tasks need some instruction-mostly verbal explanation**

UMI Examples

100	Ballpoint pen
90	Kitchen stove
80	Simple calculator
70	35 mm camera
60	Programmable Microwave Oven
50	Multi-function sewing machine
40	Multi-function digital watch
30	Complex calculator
20	Programming a home computer
10	
0	

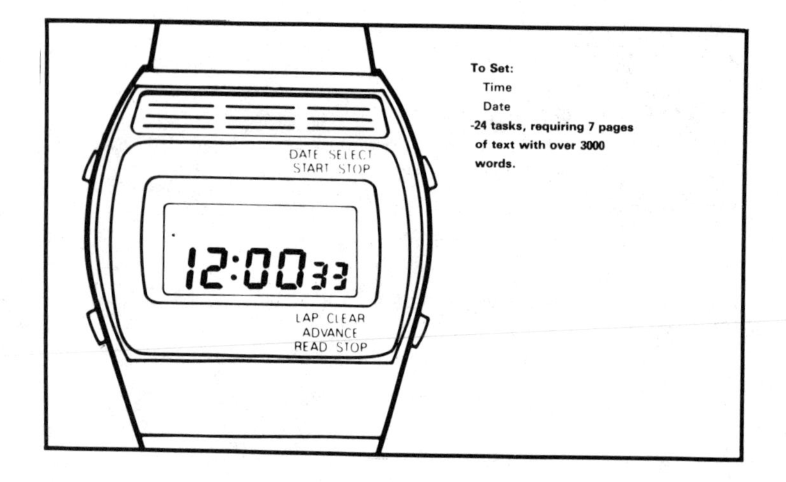

Figure 13.

A digitial watch was tested with the UMI standards. Only 43 per cent of those tested could set the watch using the instructions. Not one of the 56 participants could set it without the instructions. Of the 43 per cent who could set it with the instructions, only ten per-cent could repeat their success a second time without directions.

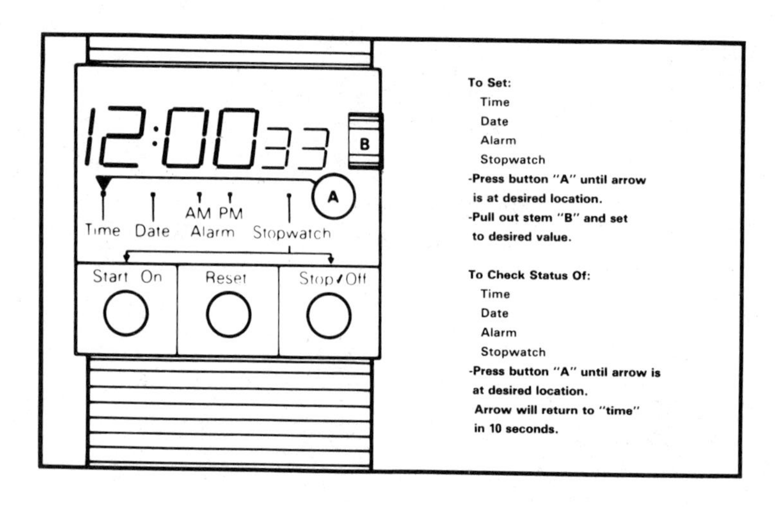

Figure 14.

The digital watch re-designed with all the necessary instructions printed to the right. The user simply presses the A button to move an arrow over whatever variable needs changing, and then makes the adjustment using a standard watch stem.

Figure 15.

The old and re-designed digital watches are plotted on a UMI graph with the old design represented by the solid line and the re-design shown as a broken line.

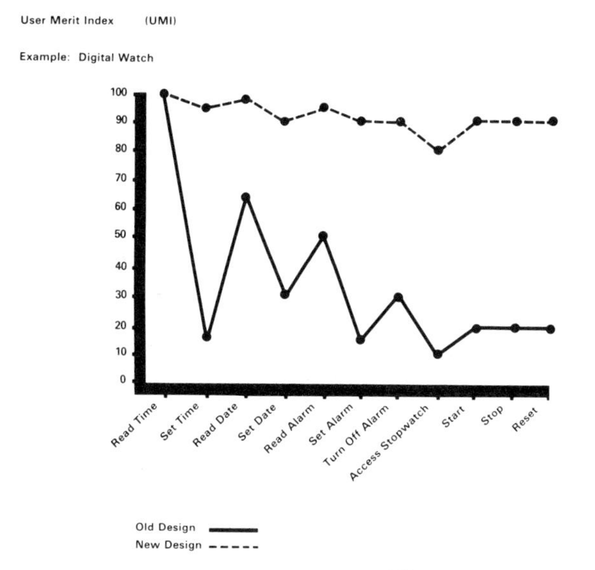

abilities in mind. They appear by and large, to have been designed by engineers who themselves have a great understanding of how the system is put together, but no idea at all as to how the user fits into that system.

We re-designed the watch to make it simpler and easier to use. (Figure 14.) A button moves an arrow over the variable the user wants to adjust. A standard watch stem sets the variable. The beauty of this design lies in its simplicity. The watch stem is a well-known stereotype for adjusting watches. People see it, recognize it, and understand it immediately. (Figure 15.)

We conducted one other test on the digital watch. We gave our 56 study participants semantic differentials - word pairings, like complex, simple or bright, dull - to see how they responded to the product before and after trying to set the watch. (Figure 16.) Not surprisingly, opinions about the watch fell quite

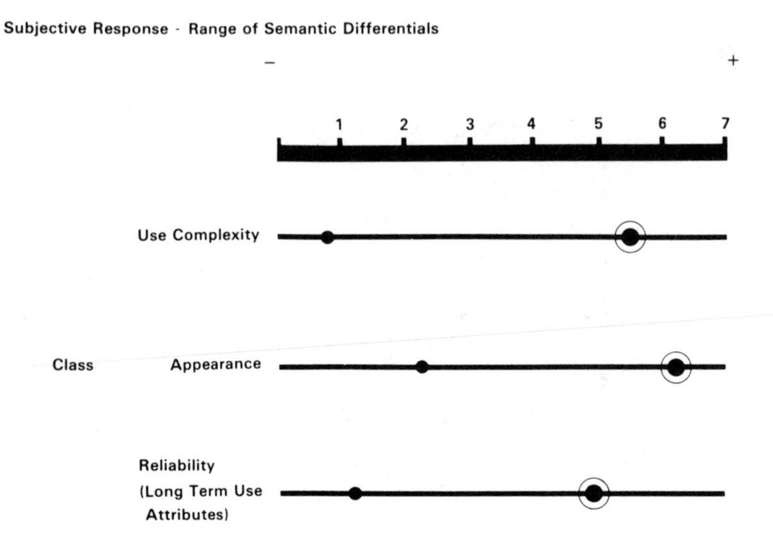

Figure 16.

A semantic differential test was used to evaluate how the 56 tested users of the digital watch responded to the product. They were asked to evaluate the watch before trying to set it, and after having tried to set it. The heavy dot with a circle indicates the initial response; the smaller dot shows their later respose.

markedly after they tried to set the watch. Before trying to use the watch, participants thought it looked simple to operate, reliable and attractive. After the test, participants said it was difficult and unreliable.

Perhaps the most telling and concise remark we heard about the digital watch came from a woman who participated in the testing and said, "I'm very upset by this design, because it's not honest. When I first looked at it, it looked very simple and easy to use. I feel now that I wouldn't buy this product or anything from this manufacturer because it's not realistic. I don't have a good feeling about this product."

This change in opinion may explain in part, why some products sell so readily for several years, and then suddenly stop selling. The manufacturer who made the watch we tested certainly suffered this phenomenon - he has dropped the watch production altogether. Although there are the factors of marketing and competition to be considered, the dismal human factors and user performance of this watch certainly figured in its quick decline.

Most manufacturers are quite good at marketing and selling their products. But matching the manufacturer's level of marketing refinement is consumers' increasing suspicion of advertising and their reliance on word-of-mouth product endorsement.

There is evidence that more manufacturers are realizing that it is not enough any longer to simply sell a product. The product must accomodate the user. An example is a new clock designed by Sony. Like the digital watch we studied, this clock uses the latest electronics and micro-processor technology. But, unlike the digital watch, this new clock is easy to set. It is clearly based on some very significant human factors research; large simple knobs allow the users to see exactly what they are doing as they set the alarm. Rather than promoting the technology that went into making the product, the company in this case, is advertising the clock as something simple to use. It may be a small sign in a welter of discouraging signals, but this product and its promotion indicates a re-evaluation of priorities, one in which the user is more important than the machinery.

The Department of Crude Arts: Viewing Videotex and Teletext from the Graphic Designer's Perspective

Aaron Marcus

Designers' talents are needed to bring order to new communication media. Those designers who approach this work will find constraints in the media's artistic refinements, and liberation in its rich informational potential.

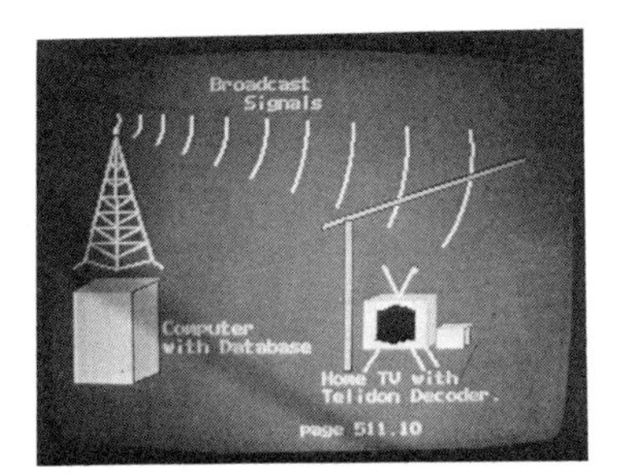

A computer-generated image showing how videotex and teletext systems work. Information frames are stored in the computer and then delivered by broadcasting signals to television sets equipped with devices capable of decoding the signal. Teletext systems allow the user to pick up only that information being transmitted. Videotex offers more choice to its users; because videotex employs either a two-way cable or a telephone line, videotex viewers may search the computer's memory and call up whatever they want at any given moment.

I sometimes wear a Mickey Mouse pin on my lapel, and people ask me why. It has a double meaning. First, I advocate the end of "Mickey Mouse graphics" in computer graphics. Much of the computer graphics made and shown in the last ten years, while being very sophisticated technologically, have not been very sophisticated in terms of their visual communication. The technology behind computers is sufficiently advanced now for attention to be turned to other important questions. For example, now that we have something on the screen, why is it there? What is it supposed to be doing? How is this communication or information being understood by the people receiving it? Questions like these, asked of this glowing computer screen, are typical of those asked by graphic designers about conventional means of creating imagery. They are the serious questions that signal the end of unsophisticated graphics, and the beginning of serious computer graphics for visual communication.

The second meaning behind my Mickey Mouse pin lies in its symbolic value. Mickey is one of the best known images in the world. One can go to almost any place, show Mickey Mouse, and people will know who he is. One of the most important things about Mickey is that he doesn't look at all like a mouse. Mickey Mouse is a very abstract, artificially constructed symbol which communicates certain things to a great many people. In that respect, he is an example of highly successful graphic design. One goal in developing better computer graphics systems is to create such potent symbols, visual metaphors, and visual communications. They must speak clearly and effectively to a large group of people and must appeal both to the emotional and intellectual structures of people's minds.

One area of computer graphics of special interest to graphic designers is videotex. Videotex, and a related system called teletext, are attracting a great deal of attention from people and corporations interested in finding ways to provide large

Two videotex work stations. Each station was equipped with two monitors. One showed the "menu" of available options - including typefaces, type sizes, and colors; the second monitor displayed the image as it was being created. Students moved the cursor around the screen with a joy stick, and entered text using the keyboard.

amounts of information to individual homes. Some videotex and teletext make use of the 20 blank lines between video frames on television sets. These blank lines are seen as a wide black band across the television screen when the picture rolls. In the early 1970's engineers realized that something could be pumped into those blank lines electronically. By fitting television sets with suitable decoding devices, it was possible to display still frames of material. Of this realization were born two approaches to delivering computer-based texts and imagery to home television sets. Through broadcasting signals, the teletext system delivers several hundred pages, or frames, of information stored in a central computer. Those images are decoded by a special device on the receiving television so the viewer can choose from among those frames being sent at that particular moment. The videotex system is similar in many respects: it too uses a type of broadcasting signal, a central computer, and is delivered to home television sets. But with videotex, the television is connected to the computer through a two-way cable or telephone system. This connection allows one to search the computer's memory at will. In this way, the viewer has access to much more information than a teletext system can provide at any given moment.

Corporate institutions now realize the potential profits of teletext and videotex. Three major corporations, CBS, Time, and AT&T, are known to be investing a great deal of money in

A graphic designer worked to improve the letter forms for Telidon, and while the product of that effort shows considerable improvement over the untouched letters, the forms are still very crude because of the low resolution of most affordable computer graphics systems.

videotex. An article dated Ocober 5, 1982 in the Wall Street Journal, reports the following: CBS and AT&T may begin a videotex business in 1983 after running a seven-month test in New Jersey. CBS has already spent about $10 million dollars on videotex. No one at AT&T will say how much has been spent there, but it is probably five to ten times as much as that spent by CBS. By the end of this decade, the article goes on to say, $10 billion in revenue are expected to be generated by videotex services.

Others are distressed by the possibility of losing profits because of videotex. Newspapers and book publishers are worried that it will replace their conventional media. At the same time, some of those publishers and journalists are intrigued by videotex's potential for transmitting information. There is evidently, a strong market for videotex: one recent article says that one of every six consumers in the United States is a serious candidate for information services of this kind. Nor is the interest confined to the United States. There are videotex systems in use in England, Germany, Japan, Canada, and in France. There is, in fact, a particularly strong French commitment to promoting videotex information services. A ten-year project in France calls for replacing everyone's telephone with a terminal, practically free of charge. Clearly there is a strong movement, and one in which designers should be involved.

I was invited by the Canadian Arts Council to go to the University of Alberta in Edmonton, Canada, to teach a workshop in the Department of Art and Design. The students and I examined the problem of creating visually literate images for a computer graphics system. For this mini-course at the university we used a videotex system. The University of Alberta's visual communication department did not have the money to buy a $50,000 work station for the class I was to conduct. Instead, arrangements were made with the telephone company in Alberta to make equipment available to us. They were conducting a videotex experiment and gave us time on two workstations, and television monitors equipped to receive images broadcast from a central data bank. At the workstations, students had choices of type sizes, eight colors, and a series of gray levels. For entering text, students used keyboards. Joy sticks were used to move the cursor and symbols around the screen.

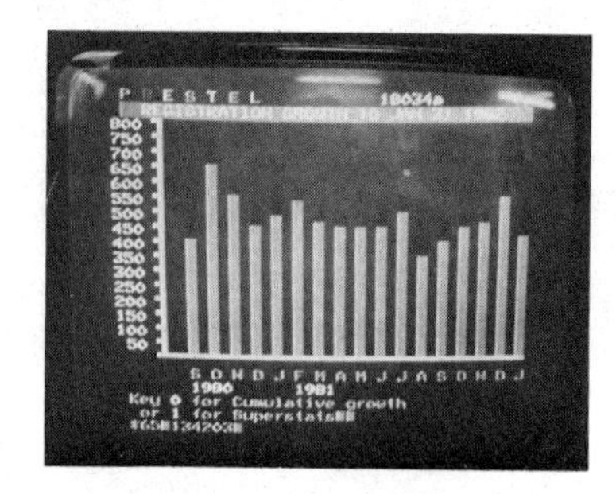

The system used to create the above image has very low resolution, yet it was still possible to create a clear, uncluttered diagram that communicates complex information in an understandable fashion.

We worked within the national Canadian videotex system, called Telidon. The Telidon system is unique in some respects. Its creators tried to allow for future development in television screens and built into Telidon the technical capability to accomodate higher resolution in better monitors.

The Telidon system will display about 40 characters across the screen and 24 lines down the screen. Considerable effort was devoted to designing enhanced graphics capabilities for Telidon. A graphic designer was employed to improve the quality of letter forms for the system. Despite that effort, a traditional typographer might still be understandably appalled by the quality of Telidon's typography.

Here we face one of the realities of working with teletext or videotex. Because the computer graphics systems available now at a reasonable cost are low resolution, one is faced with accepting a certain crudeness in type and symbolism. Those working in computer graphics have the very challenging task of designing for a mass audience with very coarse displays. Until high resolution display systems become widely available at a reasonable cost, the roughness of the media must be accepted. There is a trade-off, however, one that is not entirely to our disadvantage. While one is forced to accept a certain crudeness, a fineness is gained in other respects. With videotex in particular we have lost a fineness in spatial detail to the condition of low resolution; but we have gained high resolution in time. To put the contrast more plainly: in order to change a

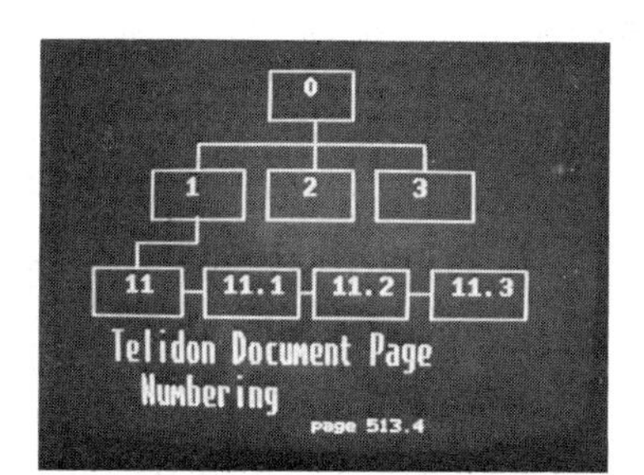

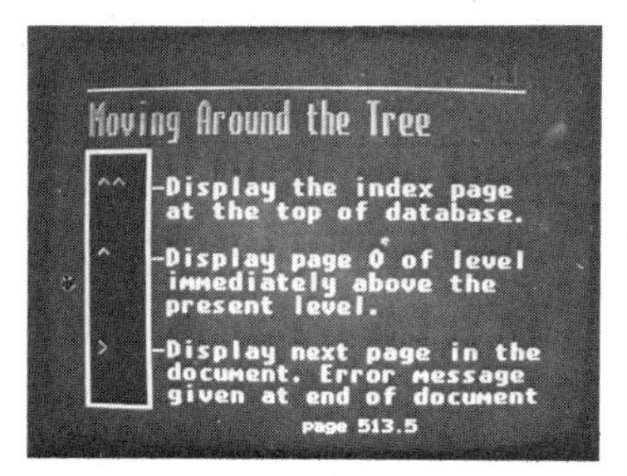

Videotex systems offer access to large amounts of information. All that information is useless, however, if the person using the system cannot find his way through the system. It is not yet a standard consideration, but as more videotex systems are created, designing clear user guides for these systems will become a major concern. Here are samples from just such a guide, designed to introduce the user to the Telidon system and the way it works.

book, one might have to wait one year until a new edition came out; in a videotex system, an electronic book can be changed instantly by altering one, or a number of frames. In working with videotex, the graphic designer, or visual communicator, is working in an interactive, dynamic media in which things can change constantly. It is a new experience for most designers.

Videotex gives access to large data bases of frames of information. The structure of the Telidon and most videotex data bases is simple. While computer scientists may sneer at their simplicity, for graphic designers, the data base of information frames is interesting and powerful. The data base provides a mass communication system in which designers can design imagery, symbols, layout and typography.

Part of what the designer must communicate to the system's user is a way to navigate through the frames of information. The videotex system is something like a book: it contains organized information. Unlike a book, it is not bound at the spine, and it does not force the reader to follow the author's sequence by turning from page one to page two, and so on. The videotex frames have a hierarchical structure that more closely resembles a tree. One can follow a series of information frames sequentially, or jump from one frame, or branch of the tree, to another.

Visualizing the way that information is structured and options by which a user can negotiate his or her passage through the system will become a significant graphic design problem in the near future. The people who have designed the current videotex systems are not always sensitive to the needs of users. There are no standards for introducing people to complex systems of information. Users are often expected to go through hierarchical menus and pass through five information frames before arriving at the information they actually want. There are millions of people who cannot read above an eighth grade level. Is it reasonable to expect them to use a system like this?

An information frame from the Telidon-based CanTel system. The graphic designers who worked on this system have taken a corporate design approach to the problem. Menu selections are placed uniformly at the bottom of the frame, and elements of the Canadian flag appear in each image as a logo.

It would seem obvious that there is a role for designers in organizing the information and designing the frames of information in a complex data base. What is remarkable is that designers have, in many cases, not been involved in such projects. In fact, strange as it may sound, some videotex and teletext services conducted consumer response tests to judge the effectiveness of their systems using information frames created by people without much background in graphic design or visual communications. Many companies are now experimenting with how to construct these frames of information using combinations of typography and imagery to introduce a system and to make the user feel comfortable with it.

The designer working on a videotex or teletext project has clear constaints: the difficulty of low resolution, a limited color palatte, equally limited typography, and the ability to create on the system only a crude sort of image. Despite these limitations, the designer can improve the use of color over that which is often seen on test systems, where every color available on the system has been employed. The designer can also form the material being presented.

There are new graphic opportunities emerging. Through careful management of the capabilities of the videotex and teletext

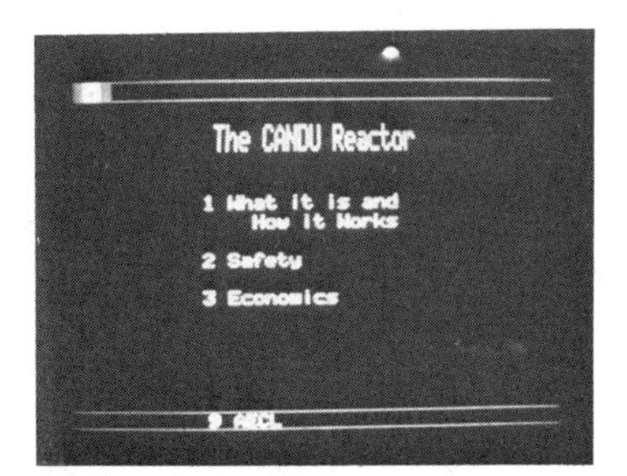

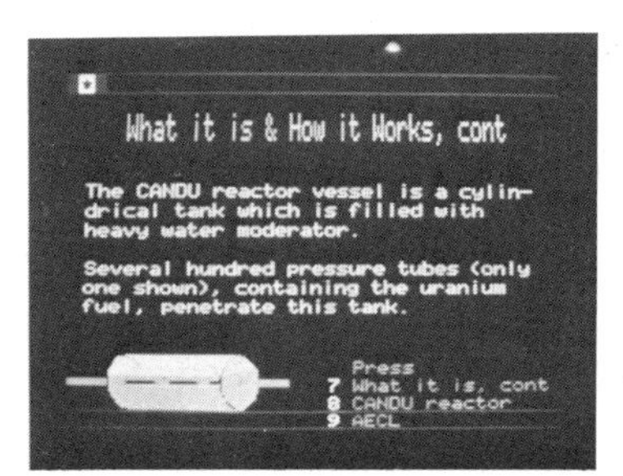

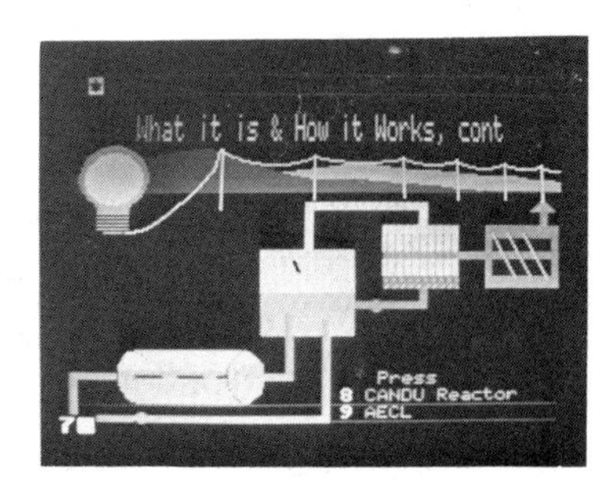

Three images from a series of information frames on the CanTel system explaining how a nuclear reactor works.

systems it is possible to make photographic-like displays. Using some of the newer work stations, a designer can focus a television camera on material. That photographed image then appears on the screen to be edited or sent directly as a frame of information. It is also possible with videotex to exchange mail with other users. Here, the graphic designer can play a role in determining how that mail will look when it comes over the system, and how a user will learn about the mail delivery facility.

A number of graphic designers have worked on CanTel, a Telidon-based Canadian videotex system. They have taken a more corporate oriented graphic design approach. Menu selections are placed uniformly at the bottom of the page. Color selections are more standardized. A logo using elements of the Canadian flag appears in each frame to remind users they are using the Canadian system. In February of 1982, there were approximately 50,000 frames of information one could have access to, and that number was growing significantly every day. The CanTel system provided among other things: an explanation of how a nuclear reactor works in a series of frames growing progressively more complex; games to play (some of them exploring consumer interest in using CanTel); rudimentary quizzes; employment information; and some experimental computer-aided instruction. Attention has been given to graphic design issues in CanTel, and that care is apparent in the more disciplined, organized use of frame space and color. Because of limitations in the use of space, information frames become a new kind of communication which mixes brief journalese with poster design in order to communicate things rapidly.

The workshop at the University of Alberta lasted only two weeks; in this short time there was a great deal to be done. I began by introducing the Department of Art and Design to computer graphics and information management through a series of lectures. Working with a smaller group of students, I introduced the Telidon system. They had only a few days to experiment with using the graphics editing capabilities of Telidon. Then they began the work of designing. They discovered the capability and flexibility of the system in terms of typography, color, and location. They also discovered its limitations. At the same time, they were reading about teletext, videotex, and information management.

I gave them the project of communicating information from the Alberta Department of Statistics through experimental information frames appropriate for use on a system like Telidon. Because information graphics often requires communicating complex material, much of the project time was spent studying, thinking, planning, and experimenting with prototypes and presentations. For these preliminary stages the students had to work with a designer's traditional tools, paper, pencils, and pens. This was done for two reasons: first, because it was fastest; second, because time at the terminals was at a premium. Unless it is possible to have a work station for each individual, time at the terminals must be allotted sparingly. Another harsh fact of life that should be borne in mind when working with a computer graphics system is the likelihood that it will not be working at least some of the time: complex computer graphics systems do break down, and repairs take time. We were fortunate to have a spare system for our project.

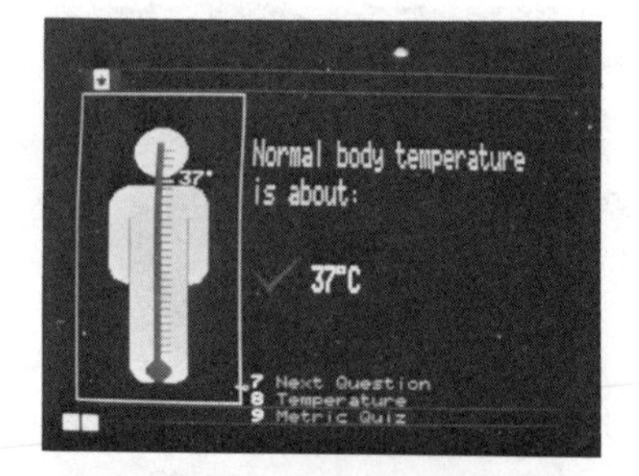

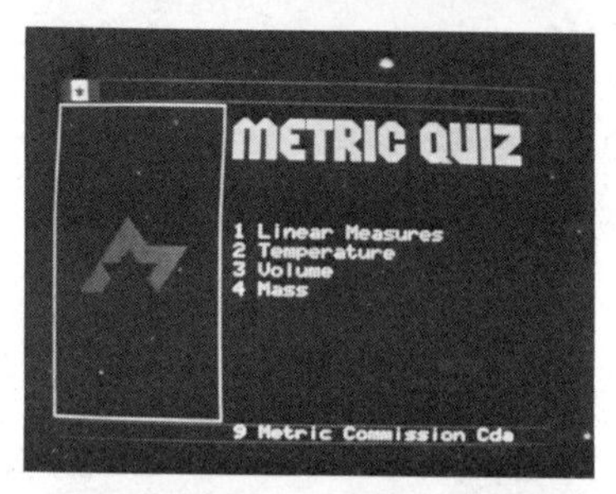

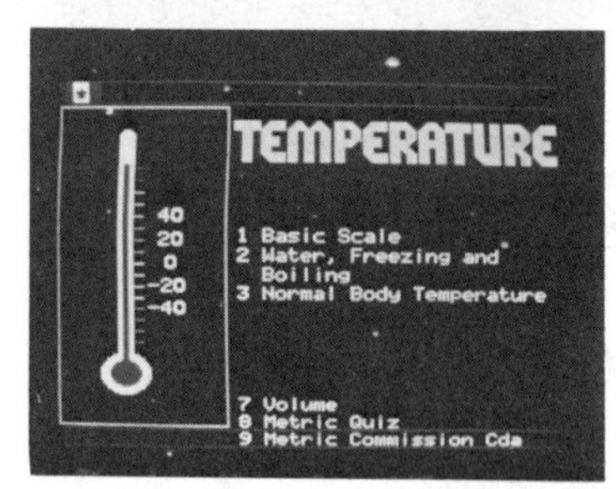

A sampling from the CanTel quiz selections.

In the early stages of the project, students worked with the traditional tools of graphic design, pencils, paper, and pens. This is because it is still a faster and more efficient way of sketching out ideas, and because time at the computer terminals was at a premium.

Without it we would have had trouble finishing anything in two weeks time.

Of their own volition, students divided the work into the component tasks and roles required by information communication: some became experts in the system; some became art directors and graphic designers; and others were researchers.

In designing the information frames, the students gravitated toward the functional graphics approach. The images they created were orderly and constrained. For large data bases it is an appropriate choice. The alternative eclectic approach would prove exhausting for the user forced to contemplate and understand all the different frame structures.

In the little time they had, the students examined a number of basic design problems. They tried to create a system logo to be used on each information frame that would work equally well at a large or small scale. In designing the information frames, the students established grids for information layout so the frames would have a consistent look, and the user would know from one frame to the next, where to look for the desired content. Color standards were outlined for the information frames. The students also tried to improve the system itself by re-designing the symbolism used for some of the command and control features, and by repairing the letter and word spacing of the system's typography. Finally, they designed some of the frames communicating information about Alberta using pictographic or ideographic means, whenever possible.

Some of the students involved in this project ended the two-week workshop feeling very excited about the potential of graphic design and computer graphics. One of them began to work in computer-aided instruction using the university's Plato system.

The structure of videotex and teletext systems demand the creation of vast numbers of information frames if these systems are to be used to their full potential. Currently, individual frames must be designed one-by-one by people. However, information frames will not be individually designed by people for long. Research is being conducted now on a semi-automatic design process that would allow computer systems operating with pre-programmed knowledge of graphic design to produce information frames. This will come about, in part, because it will be impossible to hire enough designers to create the millions of necessary frames.

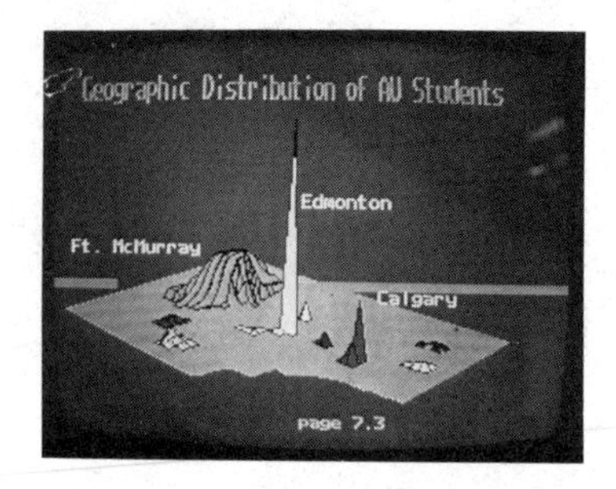

An experimental attempt at communicating complex information in a single frame using techniques borrowed from illustration and mapping.

This does not mean the need for designers in computer graphics will abate; it means the needs will change and evolve. In the computer graphics systems being created, there is and will continue to be, a great need for graphic designer's expertise in typography, spatial arrangement, and color selection. More important will be the influence designers can exert on computer manufacturers by advising them on how to produce better systems for the future. There will be a constant need for individually designed frames as well as graphic design systems for many frames. In the end it is quality that is important, even for computer graphics systems whose imagery is rather crude.

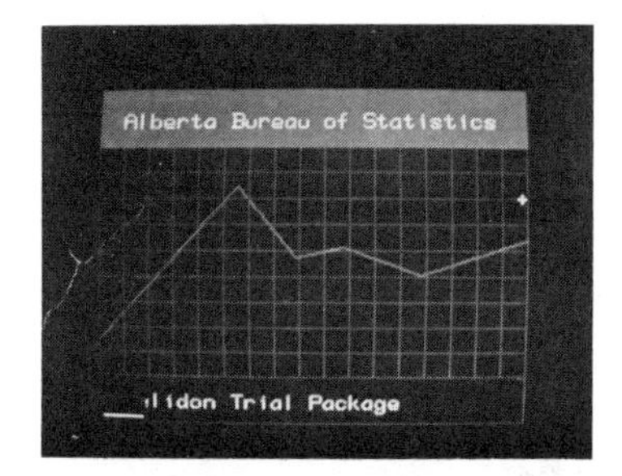

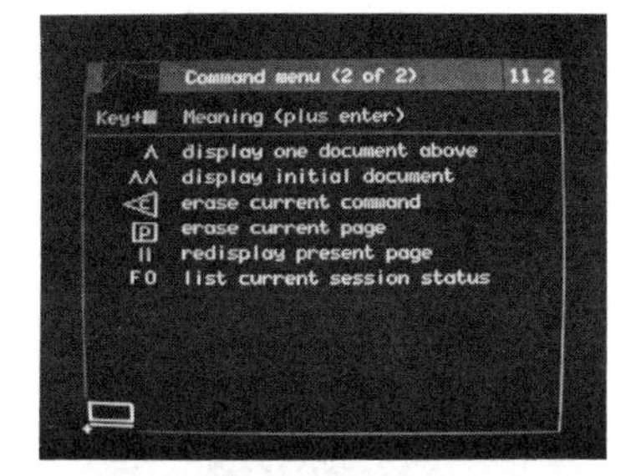

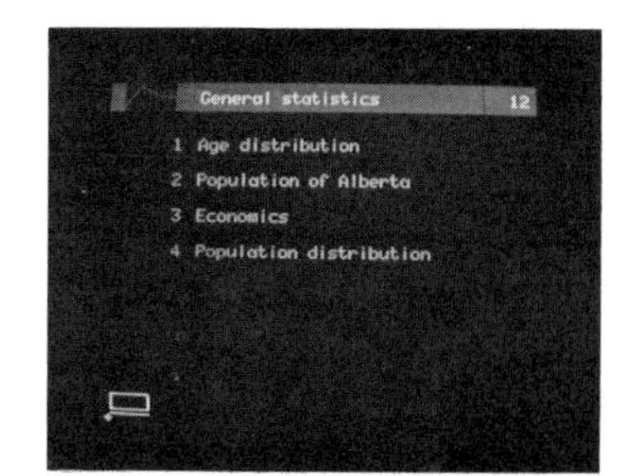

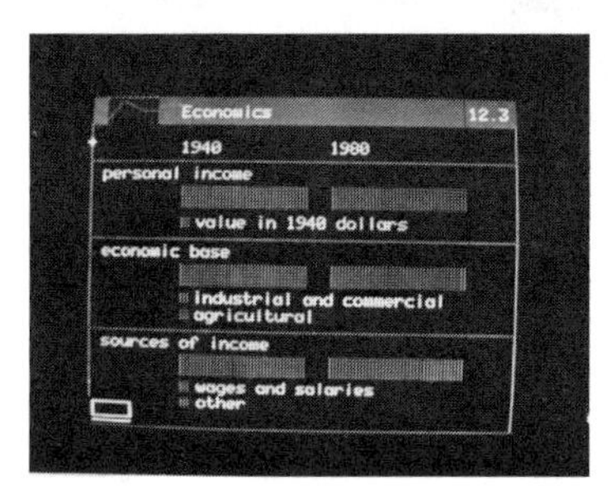

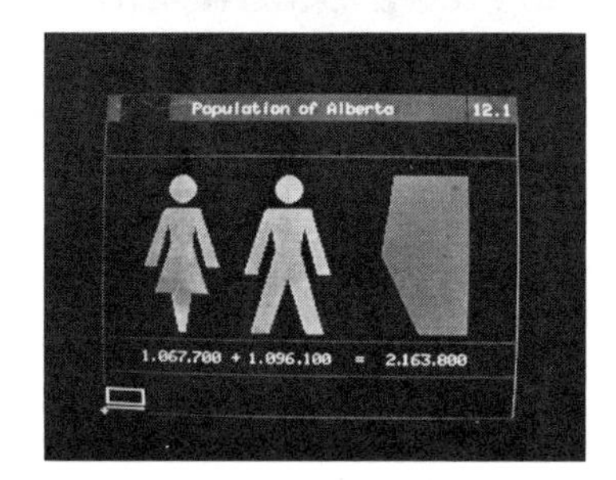

Five images created by students in the course of the workshop given at the University of Alberta in Edmonton, Canada. A system logo was designed and used on each frame of information that would work equally well at a large or small scale. Grids were established so the layout of information would have a consistent look. And, general types of information were confined to specific areas of the frame so that a user would know where to look to find text, menus, and titles from one frame to the next.

Computer Graphics and Visual Literacy

Charles Owen

The computer technology explosion, increasing sophistication in personal computers, and new developments in computer graphics are combining to set the stage for a major breakthrough in the ways we use visual language. The business world now projects a $4 billion dollar annual market by the end of 1990 for business graphics. Managers and decision makers in increasing numbers are turning to computer graphic methods for examining how their companies and institutions work, and a significant demand is growing for new and better ways to make sense out of complex data. At the same time, a new generation is growing up with computer games and personal computers. How will our notions of visual literacy benefit from the new synthesis? What can we look for from the new technology as concepts of diagramming are carried to new levels? What should designers be doing about it?

I am working now in the field of diagramming and business graphics. The specific project I'm engaged in is helping to create a sophisticated graphic system for the software development company S.P.S.S., to be used in tandem with their advanced data analysis system. The implications of this work go far beyond business graphics; reaching into the realm of visual literacy, and offering the fantastic possibility of changing the way people think and work with visual language.

To understand the potential, as well as the problems presented by diagramming, it is useful to first examine the nature of graphics systems. Jay Doblin, designer, theoretician, and a teaching colleague of mine at the Illinois Institute of Technology's Institute of Design, explains rather neatly how things go together in the graphics world (figure one). He describes an inverse relationship between the level of abstraction in the message and the contribution required of the perceiver to understand the message. At the most abstract level are writing systems. To understand them requires learning of the perceiver. The least abstract are movies, television, and photographs, or, what he calls iconic models. These iconic models are so close to reality that the perceiver does not need to possess more than his or her own experience in order to understand them. For these then, the contribution of the perceiver is quite small.

A similar relationship exists between the level of abstraction and the degree of structure in graphic systems. The general rule being: the greater the abstraction, the more structure will be needed in order to communicate and have the message understood (figure two). By this standard we would still find writing systems at the top, having the greatest level of abstraction, as well as the highest degree of structure. Falling just below writing systems is mathematics and, at the bottom, are the iconic models. While diagramming does employ some symbolism, and does have some structure, it has not reached

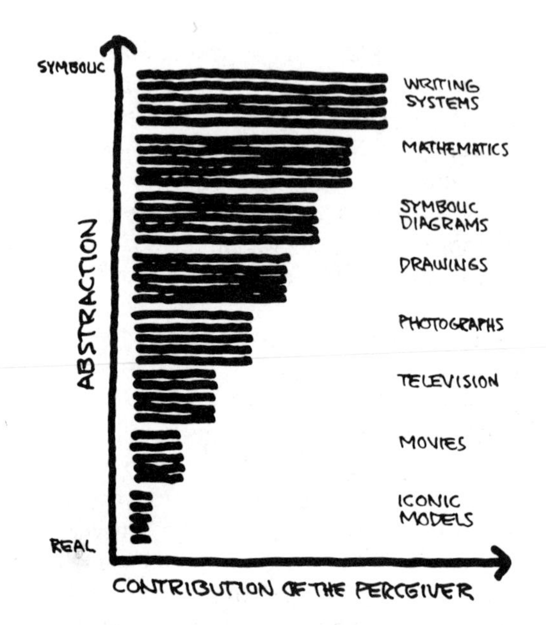

Figure one.

There is an inverse relationship between the level of abstraction in a message and the contribution required of the perceiver to understand the communication. At the most abstract level are writing systems; requiring a certain level of learning of the reader. The least abstract are movies, television, and photographs, or iconic models. These are so similar to reality that the perceiver needs only to draw on his or her own experience in order to understand them.

the level of a language. It is possible that diagramming needs more structure. It is certain that there is a potential for tremendous change in the area of diagramming.

On the scale are a number of diagramming systems that vary in sophistication and intent. Blissymbolics (figure three) is one system that is more structured than most diagramming schemes. It is capable of communicating information quite well, and can achieve something very close to the sophistication of written systems. Some work has been done on a very interesting application of Blissymbolics. Children with illnesses that impair their control over motor abilities and eye movement were taught Blissymbolics, and it appears they are able to learn communication of this sort more easily than formal writing.

Less structured than Blissymbolics, and having a highly specific use is Labanotation (figure four). Labanotation is a

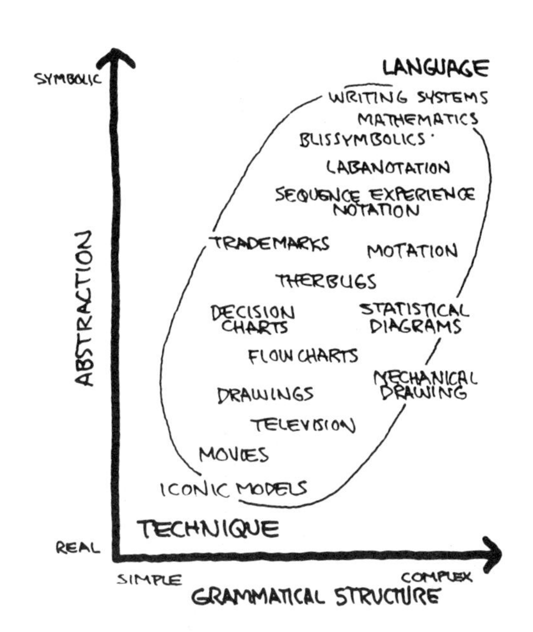

Figure two.

As the level of abstraction is heightened, the rules governing its communication become more formal. The rules of grammar and punctuation impose the highest degree of structure upon languages and writing systems. Mathematics follow with a slightly lesser degree of imposed order. In between mathematics and the iconic models at the bottom of the scale, fall a number of diagramming systems. While diagramming employs both symbolism and structure, it has not reached the level of sophistication characteristic of a language.

diagramming system that records the choreography of dances. The positions for hands, arms, feet and legs are shown in the diagrams along the lines on either side of the center line representing the torso. It is possible to achieve a high degree of subtlety with Labanotation. Using it, a dance can be recorded and reproduced with fidelity. It is extraordinary to realize that many famous, acclaimed dances have been lost irretrievably because there was no means for recording them. Even now, there are only a few people fluent enough in Labanotation to use it to record a dance well.

Sequence Experience Notation (figure five) is another diagramming system. It was designed to record a person's experience as he or she passes through an environment. It is a complex system that can be used to note very discrete items on a number of levels. It deals with time, location, objects, and activity. Sequence Experience Notation can describe, for

I RECEIVED A SAMPLE OF WHEAT FROM YOUR

AGRICULTURAL RESEARCH STATION.
(STALKS ON EARTH = FIELD)

Figure three.

Blissymbolics is one diagramming system that approaches the sophistication and flexibility of written languages. For those who know how to use it, Blissymbolics is capable of conveying fairly complex messages. Children with motor disabilities seem to find it simpler to learn to communicate with Blissymbolics than with a formal written system.

example, a person's fifteen-minute walk down a street, recording how the person was conveyed; what they did; and where they were after five minutes, after ten minutes; what that person saw, and so on. Landscape architects and environmental designers find this system particularly useful.

Motation (figure six), like Sequence Experience Notation, traces a person's path and experience as he or she goes through a designated environment. While Motation is simpler to use and understand than Sequence Experience Notation, it is also less sophisticated so that less can be communicated with it.

Another system called Therbligs (figure seven) was designed by an industrial engineer to explain how to do various tasks in manufacturing plants. The symbols can be put together to describe processes and to replace the text in explanatory charts posted in the plants.

Figure four.

Labanotation was conceived for the very specific application of recording dance choreography. The positions for hands, feet, arms, and legs are indicated by the lines coming out on either side of the center line which represents the torso. It was intended that Labanotation give choreographers the same power to record their dances as composers have in using music notation to record a score.

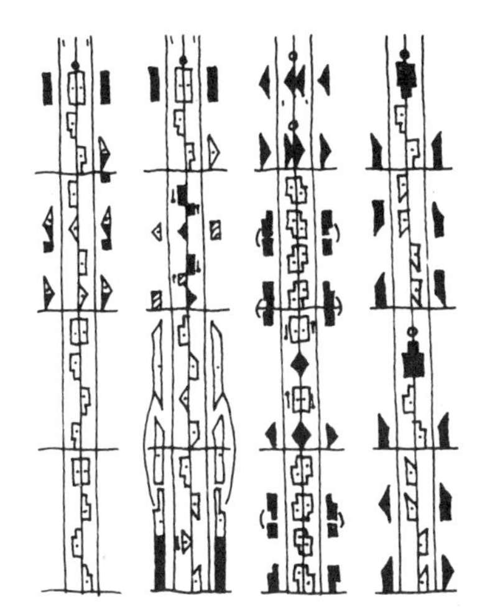

To understand diagrams in general we can analyze their most basic components - purpose and structure - beginning here with purpose.

Diagrams' primary purpose is, clearly, to communicate. What is to be communicated, and at what level of complexity, can be roughly plotted on a graph with one axis representing information, and the other showing impression (figure eight). At its highest level, the object of informing is to enlighten; in the same way, the object of impression would be to stimulate. Using these criteria it is possible to evaluate some graphic diagrams. Technical illustration is, for example highly informative, but does not make a strong impression. At the top of the impression axis are those "high potency symbols" like trademarks, logos, the swastika, and the radiation symbol. These symbols must make a strong and immediate impact, information is secondary. A low form of both impression and

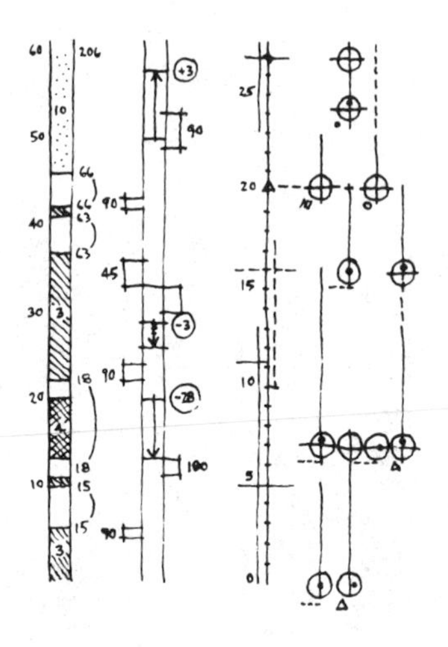

Figure five.

Sequence Experience Notation uses symbols to note and describe what people do and see as they walk through environments. It is a complex system, capable of communicating a high degree of detailed information.

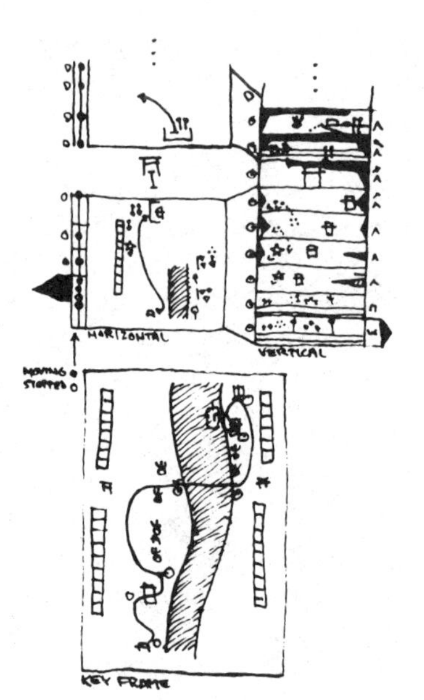

Figure six.

Motation is a simpler and less thorough diagramming system that is also used to note human behavior in designated environments.

Figure seven.

Therbligs describe processes. They are used principally in industrial environments to simplify charts by replacing explanatory texts with symbolic abstractions. A sample of Therblig characters are shown.

NAME	SYMBOL		EXPLANATION	COLOR
SEARCH	Sh		EYE TURNED	BLACK
FIND	F		EYE STRAIGHT	GREY
SELECT	Se	→	REACHING	LIGHT GREY
GROUP	G		HAND OPEN FOR GRASPING	LAKE RED
TRANSPORT LOADED	Tl		A HAND WITH SOMETHING IN IT	GREEN
POSITION	P	9	OBJECT BEING PLACED BY HAND	BLUE
ASSEMBLE	A	#	SEVERAL THINGS PUT TOGETHER	HEAVY VIOLET
USE	U	U	WORD "USE"	PURPLE
DISASSEMBLE	Da		ONE PART REMOVED	LIGHT VIOLET

information is identification. Pictograms are an example ot simple identification. A high form of impression and information is persuasion. Moving towards persuasion are: advertisements, statistical diagrams, and some business diagrams. In each of these cases someone is trying to convince other people of a need for change.

A number of people have analyzed diagramming to discover general characteristics. Ken Garland (figure nine) has pointed out that diagrams deal generally with action, force, process, or proposition. Occasionally they deal also with function and construction. That is a fair analysis, but it could go much further.

Matthew Murgio (figure ten) looks at relationships to be diagrammed and offers this breakdown: quantitative, methodological, functional and structural. Murgio has matched

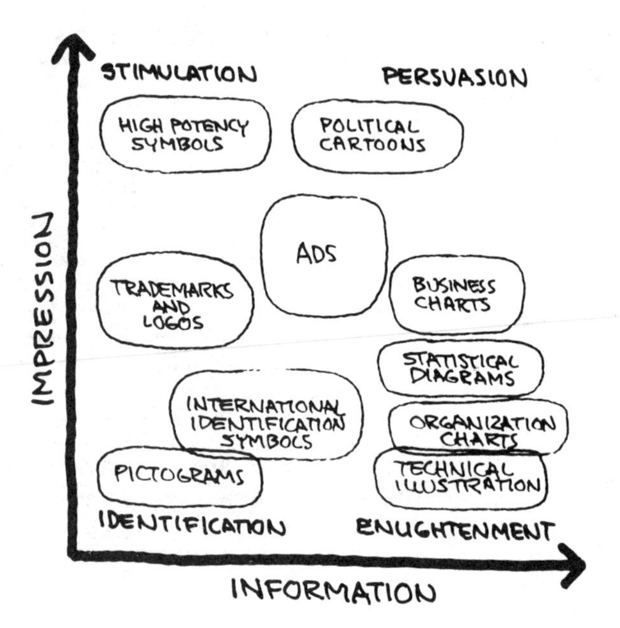

Figure eight.

Diagrams may be evaluated on the basis of their power to either inform or impress. Enlightenment is the ultimate purpose of information, while stimulation is the highest goal of impression. Applying these criteria and standards, one car roughly plot various methods for communication.

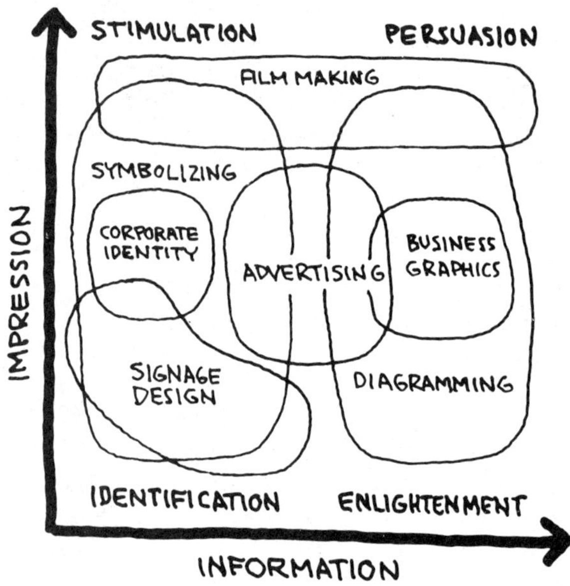

Figure nine.

A summary analysis of what diagrams accomplish.

Figure ten.

Murgio's breakdown of the types of relationships that are diagrammed, matched to their most appropriate diagramming techniques.

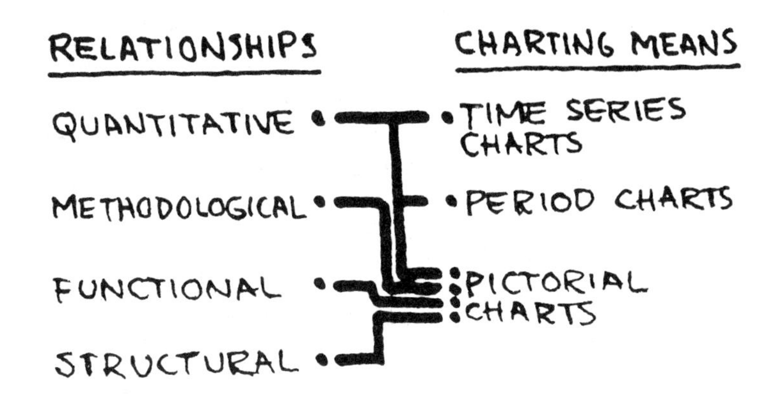

diagramming techniques to each relationship. Quantitative relationships can be diagrammed with time series, period, or pictorial charts. For methodological, functional, and structural relationships, Murgio relies only on pictorial diagrams. (Pictorial charts are, it seems, the most useful means of diagramming, and the least well defined.)

Karl Karsten (figure 11), another to have made a study of diagramming, divides diagrams under the two general headings of mathematical, and non-mathematical. Under mathematical fall, special analysis, rate of change, and amount of change. Those items under non-mathematical are more interesting; these include, space relations, idea relations, time relations and composites (figures 12-14).

The Integrated Software Systems Corporation (ISSCO) makes computer graphics programs. In the manuals that describe how

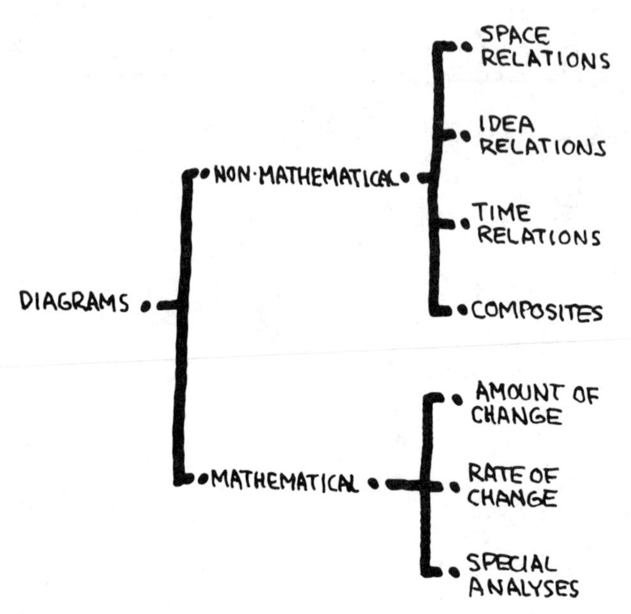

Figure 11.

Karsten's general headings for types of diagrams.

to create business charts using computer graphics, the problems to be diagrammed are divided into: time series, parts of a whole, comparisons and relationships (figure 15). The manual then suggests a solution, using a matrix to match the type of problem to the appropriate diagramming technique offered. Among the techniques available are: curves, bar charts, surface charts, column graphs, and maps. If the problem to be diagrammed is not too complex, the matching is simple, and the diagram achieved can be reasonably good, or quite good.

The person who has, I believe, gone furthest in analyzing the purpose and giving structure to diagramming, is William Bowman (figure 16). The first step in his analysis is to ask the question, "What do we want to do with the diagram?" The diagram may need to resolve questions of quantity, location, or process; or, questions of how much, where, or how. The next step is to pair an appropriate technique to the diagramming

Figures 12-14.

Samples of three non-mathematical diagrams, as defined by Karsten.

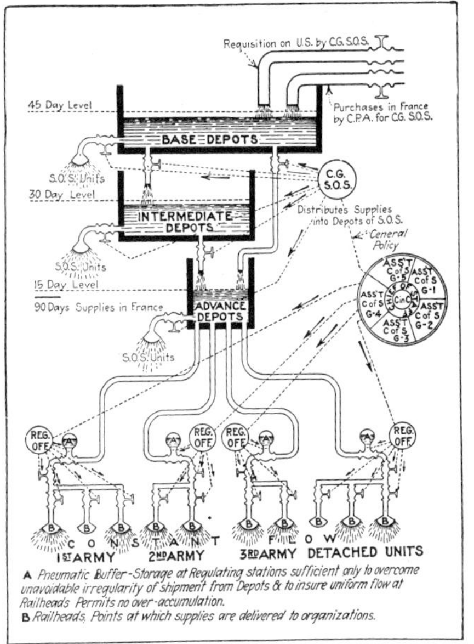

Permission of Mr. Malcolm C. Rorty.

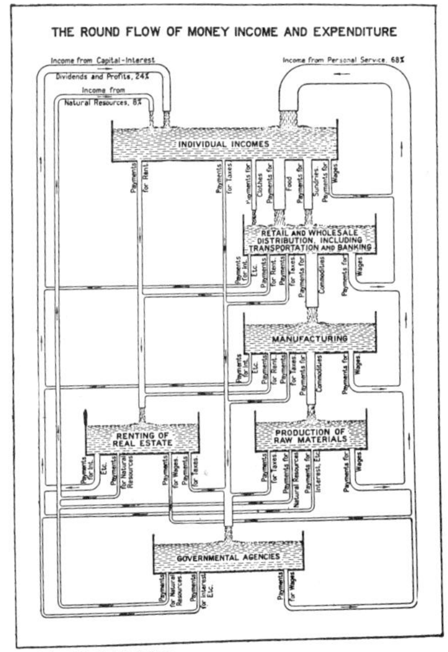

Permission of Mr. Malcolm C. Rorty.

From Bartholomew's Atlas.

problem. What makes Bowman's work exceptional however, is his attempt to extend the process further. Rather than thinking of diagramming as a simple collection of methods to be coupled with needs, Bowman thinks of the diagramming process as being broken into elements that can be assembled with some sort of structure. Bowman is pushing diagramming towards the sophistication of a language (figure 17). He talks about vocabulary, grammar, and phrasing. His vocabulary is one of form, dealing with point, line, shape, color value and texture. He describes a spatial grammar making use of, association, relative size, superimposition, and so on. With phrasing, the issue of style is introduced and addressed by means of texture, and spatial relationships. In his book which is now, unfortunately, out of print, Bowman illustrates with diagrams each of his points. There is a great deal to be learned from his work.

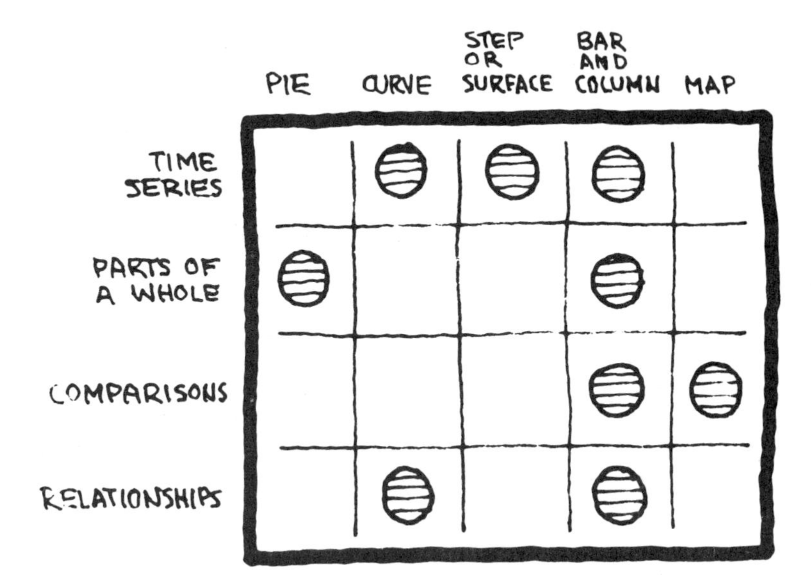

Figure 15.

The Integrated Software Systems Corporation matches typical diagramming problems to the most appropriate diagramming technique.

Bowman begins to address not just purpose in diagrams, but structure. If, however, structure in diagramming is to be made formal enough so that it can be programmed for computers, the concepts that can be diagrammed and the methods by which they are diagrammed must be clearly defined.

One useful way to understand how to structure the presentation of information with diagrams is to understand how information is interpreted best. Jay Doblin says there are two ways information is communicated, these can be characterized as presentational and sequential (figure 18). Some information must be presented in a set order or the receiver will be unable to understand it; this identifies a sequential system. Mathematical systems and written languages are examples of sequentially organized information. Presentational information, on the other hand, does not require the receiver to examine it in any particular order to understand it. Paintings are good examples

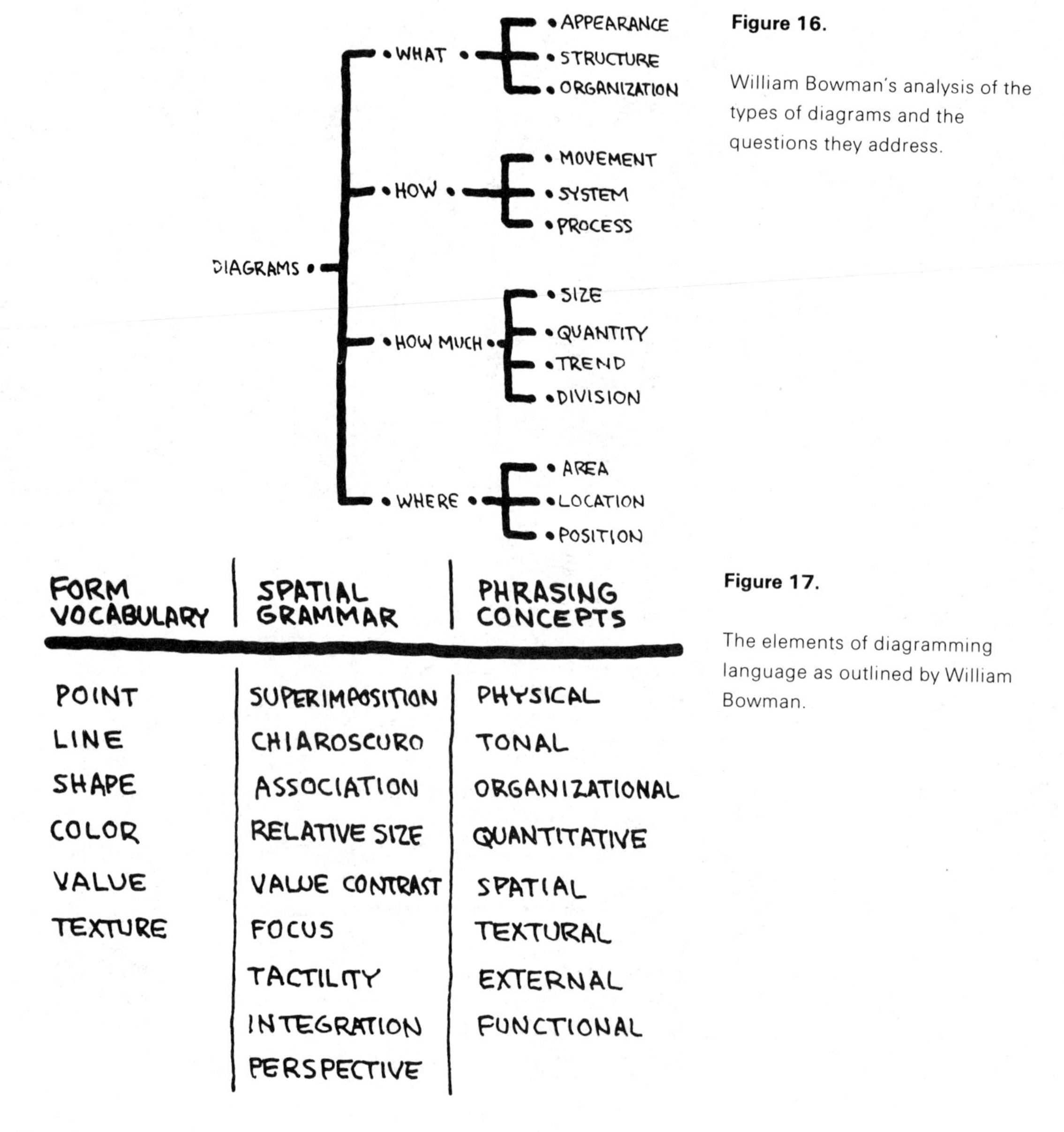

Figure 16.

William Bowman's analysis of the types of diagrams and the questions they address.

Figure 17.

The elements of diagramming language as outlined by William Bowman.

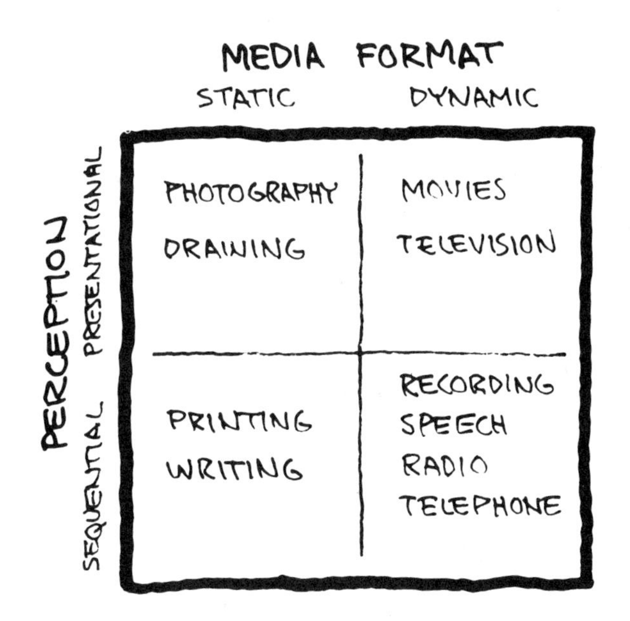

Figure 18.

An analysis of how information is communicated and understood as postulated by Jay Doblin. The media, or means used to convey the message, may be either static or dynamic. The receiver of the message may need to get the message in a logical, sequential fashion in order to make sense of it; or, the order of the presentation may be inconsequential to communicating the information effectively. With these criteria, a number of media have been plotted; radio, for example, is sequential/dynamic, while photography is presentational/static.

of a presentational graphic medium. There is order in the painting, but the person looking at it can let his eyes move over the canvas as he likes, and still understand its content.

Another way of imposing structure can be created by fashioning a kit of parts, the components of which are borrowed from systems engineering, and systems design. The parts of the kit include: contexts, entities, attributes and relations.

Diagrammatic context can be defined as space, time, or what could be called domain (figure 19). An example of domain could be found in organization charts where neither time nor space are key factors; the context of the organization chart is the organization itself (figure 20). Flow charts and link analysis diagrams are securely located in the time context. Contour maps would be defined as diagrams of space.

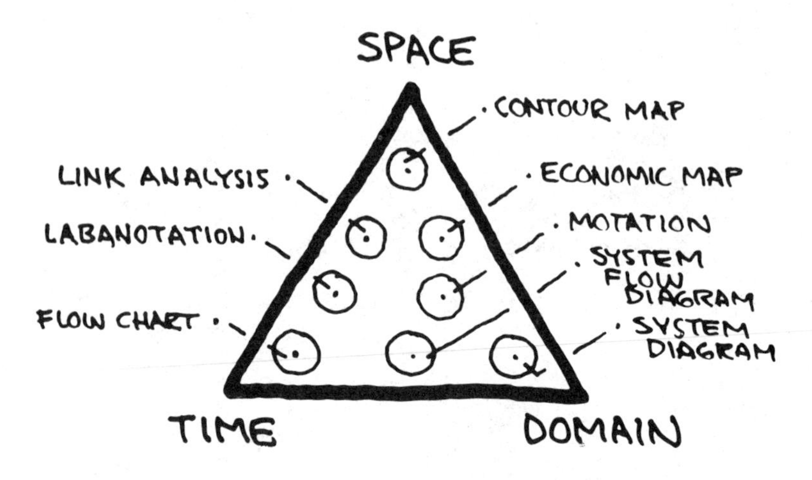

Figure 19.

Various types of diagrams are plotted here to indicate where their main purpose lies, whether it is in the area of space, time, or domain.

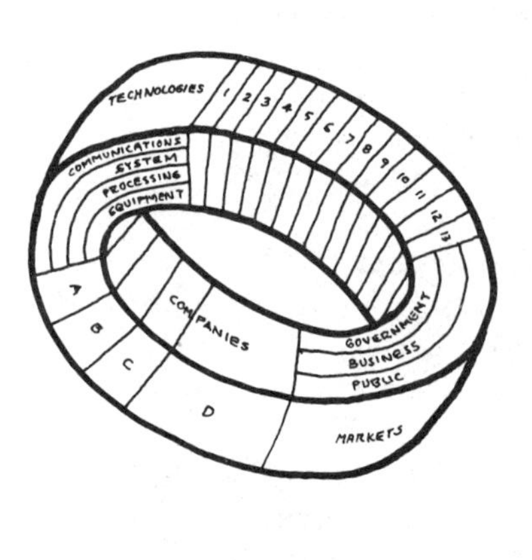

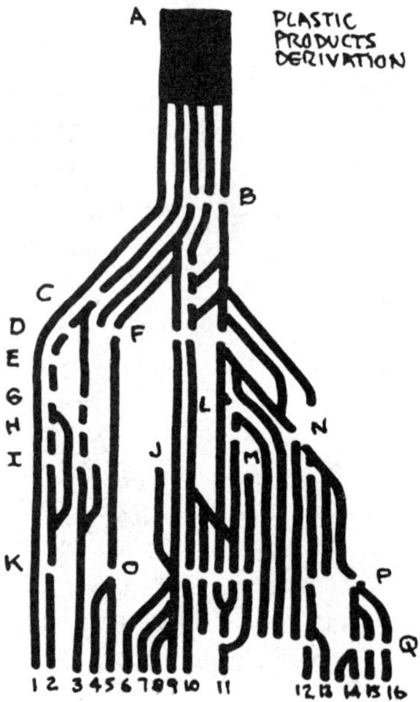

Figure 20.

Two rather extraordinary examples of organization charts. These would be classified as domain diagrams because they are not concerned with either time or space; their context is the organization itself.

After Richard Hess/Hiroshi Morishima

After Paul Bruhwiler/Saul Bass and Assoc.

Figure 21.

A system diagram.

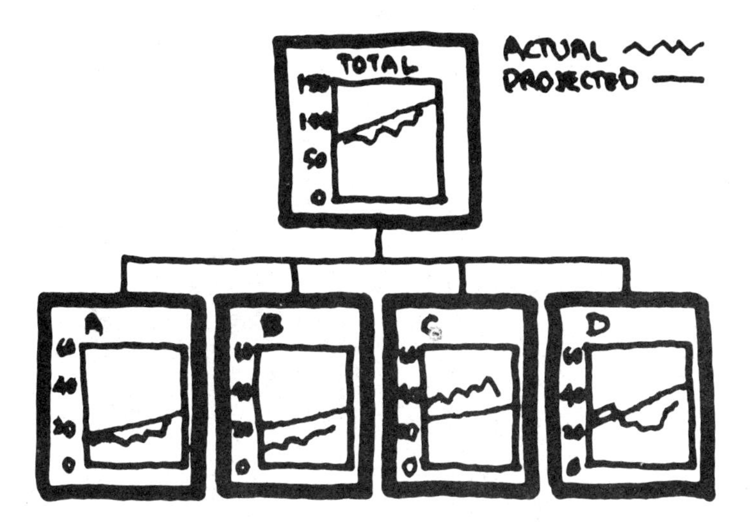

The description of entities embraces the symbolic, analogic, and iconic, or that passing from the highly abstract to the relatively real. As an example we can look at a system diagram (figure 21). The entities are analogical in the sense that the boxes in the diagram are analogs of, or represent, system elements. Bar charts are symbolic (figure 22). Link analysis is iconic because one tries to make the components of the space look exactly as they would in reality (figure 23).

Attributes are categorized by the nature of their measurements: discrete, rank order, and continuous. Discrete things, for example, are not ordered in a hierarchical manner, they are simply classified. One could classify people by the color of their eyes, dividing them into brown, green and blue categories, without implying order. An example of rank order would be music notation; an order is followed, but it is not continuous. Continuous means, quite specifically, a real continuous means

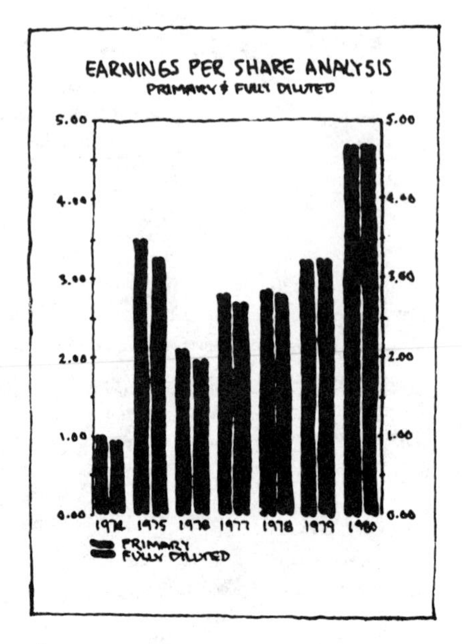

Figure 22.

An example of a bar chart.

of measurement. An example of that could be found in an X, Y plot where two attributes are plotted against each other and any subtle change can be recorded.

The various types of relationships can be described as organizational, procedural, and spatial. Illustrating an organization relation (figure 24) is a diagram showing how a French school operates. The links between each of the elements detail the organizational relationships. Music notation crops up again, here, as an example employing procedural relationships. As a spatial relation, I have chosen Labanotation as an illustration of a diagram in which relations are both spatial and procedural.

To invent a diagram, one can go through the four categories of context, entities, attributes, and relations, testing to see which categories apply, and how they fit the diagramming problem to be solved.

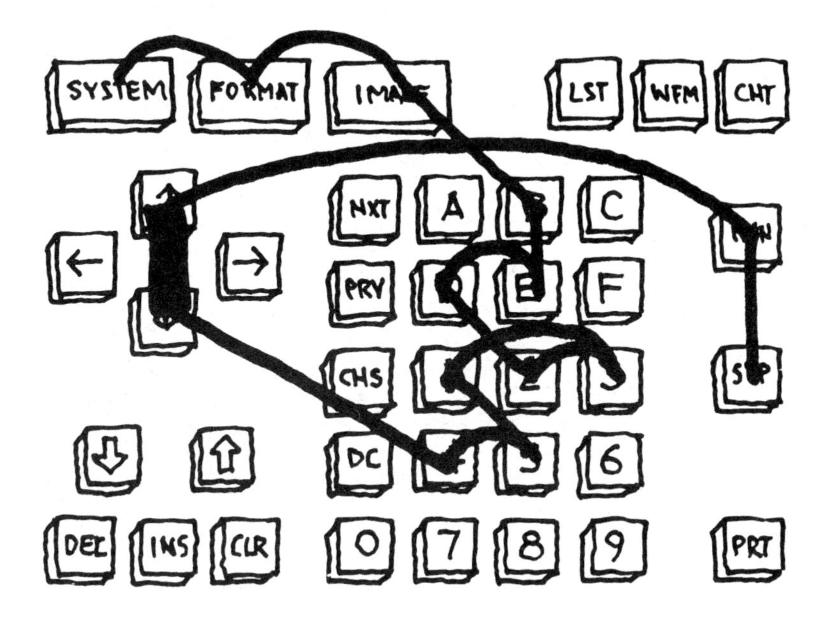

Figure 23.

An example of link analysis.

Moving to a higher level of structure, we begin to deal with diagrams with, what I call, operations. It is useful here to look at Jay Doblin's "Map of the Media" (figure 18), which divides media formats into static and dynamic, and separates their perception by sequential or presentational. To look at some examples of each type, we would find that photography and drawing are static presentational media; printing and writing are static sequential; movies and television fall under dynamic presentational; while recording, and telephones are dynamic sequential.

Diagrams have traditionally been static media forms. The first diagrams for businesses were developed in the Industrial Revolution when there was a need, for the first time, to track time and production. Diagramming is a relatively new phenomenon, one that has been confined to the page and linked to printing and other graphics media. When diagrams begin to

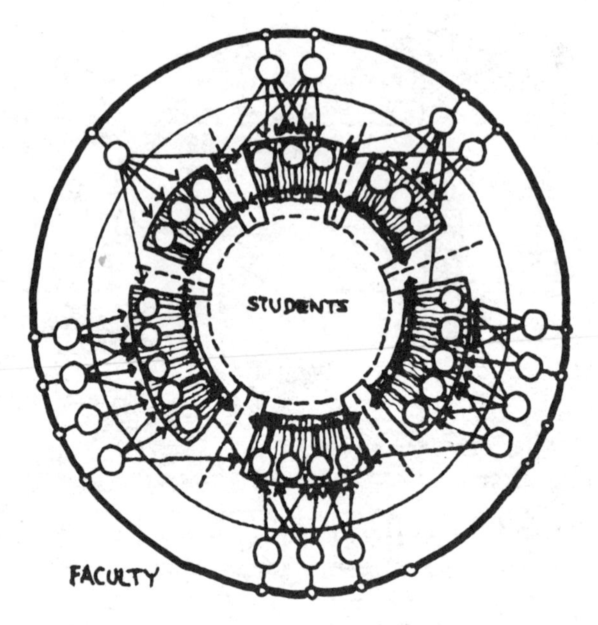

Figure 24.

This diagram of a French school's organization and operation illustrates an organization relation.

move from the static field into the dynamic area, some interesting things begin to happen, animation begins to be used, for example. Animation has not been explored to any serious degree so far, but some interesting work has been done in educational films. This dynamic potential in diagramming can be creatively exploited using computer graphics.

When we work with diagrams in real time, we "perform operations" upon them. Operations are now the most interesting of the structuring elements because they allow us to use real time to investigate and understand phenomena. They may be employed with both context and form; I call those that work with contexts transpositions, and those that change form transformations. What is common to both is that they enable us to change the way a display proceeds in time so that we can look for patterns under precise control.

Figure 25.

An image from the Lawrence Livermore Laboratory simulation demonstrating the consequences of dam failure and showing how the wave of water would move and at what rate.

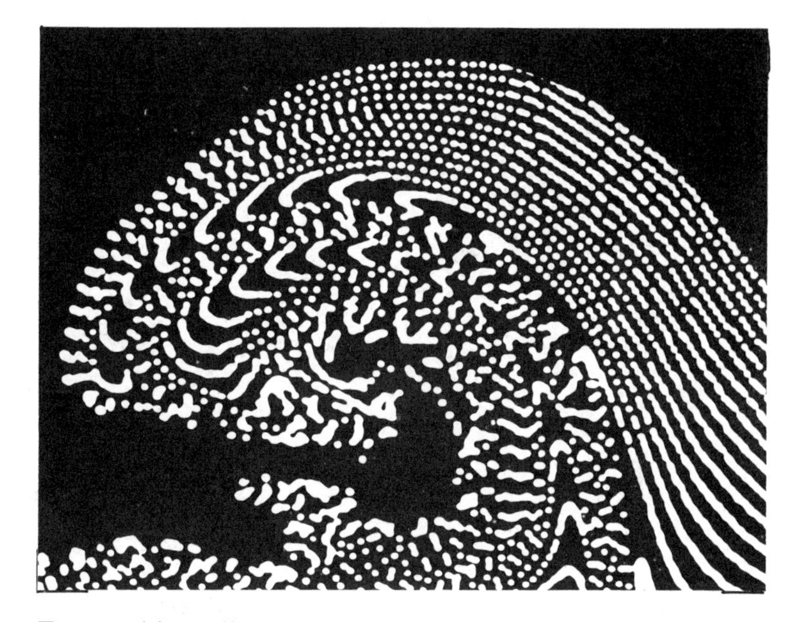

Transpositions allow us to control our views of time, space or domain. Edgerton's classic photograph freezing a drop of water splashing at the bottom of a sink is an example as are the many techniques of television sportscasting that give us freeze frames, instant replays, slow motion, zooms, pans, split screens and much more. The ability to study relationships from anywhere in space, time or domain is the operational concept of transposition.

While transpositions allow us to view a process analytically, transformations let us change it. Associated with transformation is the idea of simulation; instead of documenting the model's performance, the model is changed to see what will happen. Here we encounter again the four elements that can be controlled in making diagrams: entity, attribute, relation and context. Any or all can be altered in the process of simulation. A simulation study performed several

Figure 26.

An example of transformation. Two in a series of computer-generated maps of Chicago, highlighting areas of the city for different attributes. In one instance, poverty zones of the elderly are shown, while another map shows the distribution of the unemployed.

years ago at the Lawrence Livermore Laboratory demonstrates the concept nicely (figure 25). With a computer program, investigators were able to simulate graphically the effects of various kinds of dam failure, calculating second-by-second the positions of water droplets as the water spilled from behind the dam. Quite a different example of transformation is a computer-drawn map of Chicago diagramming in color the nature of the city for different attributes: in one instance the map reveals zones of poverty for the elderly; in another the distribution of the unemployed is shown (figure 26).

What does all this mean? The ultimate impact will be to change our notions of literacy. The capability to work fluently with computer graphic representations of data will significantly change the way we communicate and think about our world. The change is inevitable because computers are becoming more capable and less costly all the time - and their use is becoming more pervasive. We have only to know that plans in Japan call for placing 10 million transistor-like devices on a single chip by 1990 to believe the trend will continue. Moving apace with the advances in general computing is the growing sophistication of home computers. These are no longer toys, but complex and able pieces of equipment. As our power to visualize with them increases and our children learn to use them, a new era for visual literacy will dawn.

The key to the revolution is the phrase "what if?" What we need is for people to be able to ask that question and answer it by turning data into diagrams, moving them in time and space, manipulating variables and then seeing what happens. When that can be done we will have taken a giant step toward realizing the potential of visual langauge.

Interactive Documents

Steven Feiner

The powerful personal computers of the future will be capable of supplying the information we now get from books and other sources. By combining the best features of the old print media with the computer's greatest strengths, the developers of a new software system seek to solve the problems of information generation, storage, and retrieval in the new computers.

Acknowledgements: Past and present members of the Brown University Computer Graphics group who made this work possible include: Kurt Fleischer, Steve Hanson, Alex Kass, Imre Kovacs, Sanyi Nagy, Joe Pato, Randy Pauch, Will Poole, Joel Reiser, David Salesin, Adam Seidman, Charlie Tompkins, Barry Trent, Mark Vickers, Gerry Weil, and Nicole Yankelovich.

The project I am going to describe is one that we refer to as Interactive Graphic Documents. My colleagues and I believe that in the next ten to fifteen years powerful, portable, personal computers will provide much of the information that we currently obtain from books. In our research we are exploring ways of creating and presenting this sort of information. The computers that we have in mind will have a high resolution color display at least as good as those available on today's most expensive research machines, but will cost only a tiny fraction of their price - less than what a color television costs today. Their software, though internally complex, will provide a simple, straightforward interface for the user, who will point at, draw on, and even talk to pictures and text on the display.

One thing we are trying to do with our Interactive Graphic Documents is to retain the best features of the old media. Books, for example, have some wonderful qualities. One can pick up a book, turn to the place at which one has left a bookmark, and immediately get a feel of how much has been read and how much there is left to read. At the same time, we would like to eliminate some of the weaknesses of the old media. One book is a very nice thing to hold, but a library is a little difficult to carry around. With the ever-increasing amount of information needed today, it would be nice to have access to large amounts of information without having to thumb through cards in a catalogue or walk through stacks in a library. In trying to retain the good features and eliminate the weaknesses of the current media, we would like to exploit the novel features of the new computer medium.

Let's look at some of the potential advantages of the computer. One of the computer's great potential strengths is the possibility of hiding both size and complexity. We can dream of creating a large network of information that is exceedingly rich in structure, but which can be presented to a person in a clear, straightforward manner. Another strength is the ability to display high quality color illustration, typography, and dynamic, moving displays. Computers also offer the potential for give-and-take interaction in a way that a book or movie cannot: the information being presented can change in a lively and responsive fashion based on what the user does. General purpose power is another plus: there is a wide variety of tasks that the computer can help us solve, given the right software. Our computer medium can be a personalized one too. A book is typically written for many readers, and every person who reads it gets the same information in the same format. In contrast, computers can make it possible to vary the information or the way in which it is presented to suit the needs of a particular person or situation. Yet another advantage is high speed information distribution. New information can be added or old information updated instantly without waiting for a typesetter or the mails.

In our work, we have borrowed some familiar concepts from the traditional presentational media to structure information. We start with a "page", containing the illustrations and text one might find in a conventional book, but displayed on a color screen. Unlike conventional books, our pages have buttons that are sections of the page. These buttons, when touched, each cause something to happen. For example, one might run an animation sequence; another might cause a new page to be displayed. Pages have indexing information that make it possible for a person to retrieve them by using key words or phrases. Pages may also be associated with programs that will run when they are displayed.

Our pages are connected together, but unlike a conventional book there is no spine and no linear sequence to the pages. In our documents each page may have any number of succeeding pages, and any number of preceding pages. We call the connections between these pages "links". The user follows the links to go from one page to another, often by touching buttons that cause the new pages to be displayed.

Another concept we have borrowed from conventional books is that of "chapters". Instead of lumping all our pages together, we allow the author to group related pages together and nest them inside of chapters. Chapters may, in turn, be nested within other chapters. Together the pages, chapters, and links make a "document".

Given this model of storing and structuring information, we have developed three programs that allow designers, authors, and readers to create and interact with our documents. The Picture Layout System is used by designers to make pictures containing illustrations and text. The Document Layout System is used by designers and authors to lay out pages containing these pictures, to place the pages in chapters, and to link them together. The Document Presentation System is used by readers to interact with the document. The images shown here were photographed directly from the screen during the sessions with each of these systems.

The Picture Layout System (figure one) is used to lay out an individual picture from simple geometric primitives such as lines, arcs, and polygons, pieces of text, and copies of previously created and stored pictures. Pictures may be edited, permitting objects that have already been placed to be moved, copied, adjusted, or deleted. In the Picture Layout System, as in our other programs, the designer points to things on the screen using a hand held "puck" whose position is displayed on the screen as a white arrow. The designer depresses a button on the puck to "pick" or select whatever the arrow is pointing at. Here, a position is being picked for a line of text that will be typed in at the keyboard.

Our Document Layout System lets its users place pictures on pages and nest pages and chapters inside other chapters. The

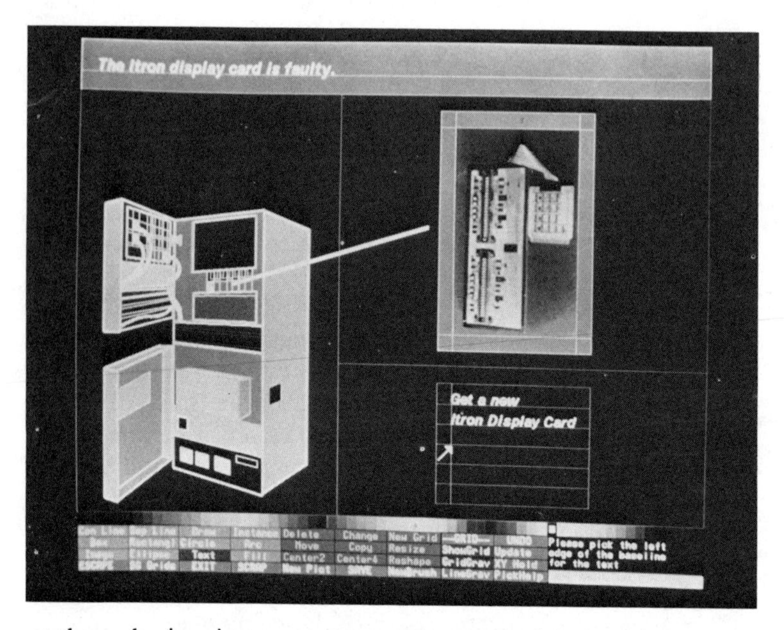

Figure one.

The Picture Layout System is used to compose pictures from copies of previously created and stored images, pieces of text, and simple geometric primitives. In this instance, the machine shown on the left is a retrieved image, and the circuit board on the right is a bit-mapped image. The arrow in the lower right of the screen shows where the user is pointing. Here, a position has been selected for a line of text to be entered with a keyboard.

author who is using our system sees a screen that contains chapters, pages, and links created during this or previous sessions. (Figure two.) Here we are looking at a chapter called "CNC M&R", an abbreviation for "Computer Numerical Controller Maintenance and Repair". The chapter is the large rectangle that takes up most of the upper part of the screen. Its name appears at the upper left. This chapter contains two pages that are presented as white bordered rectangles and one sub-chapter that is displayed as a black bordered rectangle. Links between the chapters are shown as arrows. The author is preparing to create a new chapter, has picked the buttons at the bottom of the screen labeled "Create", and "Chapter", and is about to pick the position for one of the new chapter's corners. The next image (figure three) shows the completed chapter after the author picked another corner and typed in its name, in this case "Parts Catalog". Now the author is free to create pages or chapters within this or any other chapter, to place on the pages

Figure two.

The Document Layout System Can be used to place pictures on pages, and to nest pages inside chapters. A chapter is represented by the black-bordered rectangle, and two white-bordered pages are shown above it. The links between them appear as arrows.

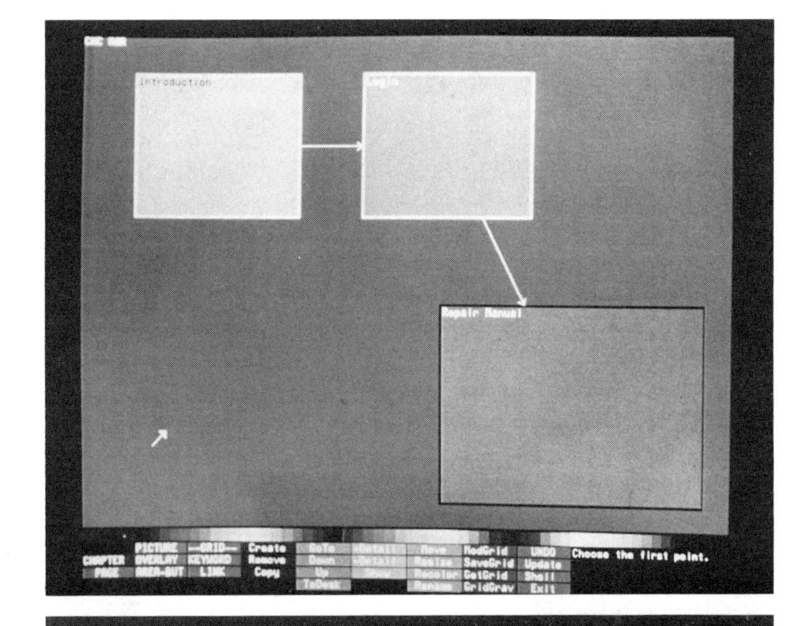

Figure three.

The Parts Catalog has been added, and links have been made to it from other chapters and pages. The author may add more pages or chapters to this chapter or nest it within another chapter.

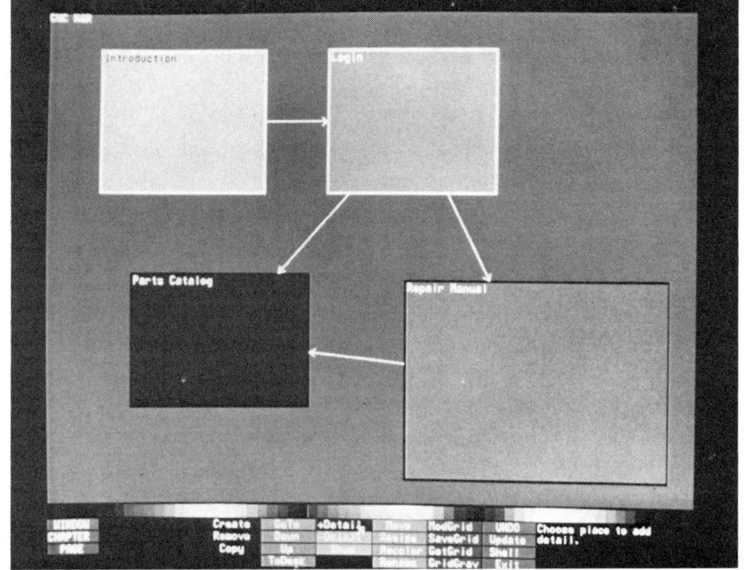

pictures made with the Picture Layout System, to make parts of a page into buttons, or to make the links that connect the document together.

In this case, after creating a few more links, the author has decided to increase the amount of detail being displayed by asking the system to show miniatures of the pictures that were already on the pages and the chapters and pages that were already inside of the "Repair Manual" chapter. (Figure four.) Since the screen is beginning to get crowded with small images, the author has now decided to zoom in for a closer look by pressing the "Down" button and pointing at one of the pages or chapters, in this case the "Repair Manual" chapter. This chapter expands to take up the workspace at the top of the screen (figure five). The author may continue to "thumb through" the document by pressing the "Down" button, pressing the "Up" button to return to a page or chapter's parent, or using other buttons that allow pages and chapters to be retrieved by name.

It is often helpful to be able to look at different, perhaps widely separated, parts of the document at the same time. If the author selects the "To Desk" button, then the workspace is filled with a set of "windows", each of which contains a page or chapter. (Figure six.) These windows were created and positioned on the desk by the author earlier in this session or during a previous one. In the preceding images the author had been working in the bottom middle window, which had been expanded to take up the workspace. The author can now move around in and alter the contents of each window individually, or make links between pages and chapters in different windows. Authors typically use one window for a document overview and show important pages and chapters, with as much detail as needed, in the others.

The document created by our authors and designers is

Figure four.

Additional detail is displayed after the author touches the "+ detail" button and points to the introduction and Login pages and the Repair Manual chapter.

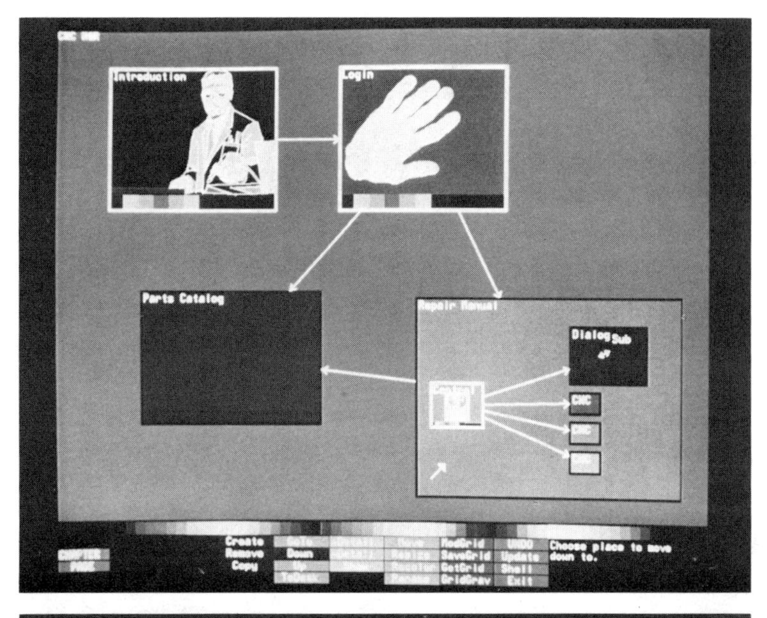

Figure five.

The author, wanting a more detailed look at one of the chapters, has pressed the "Dawn" button and the selected chapter has expanded to fill the screen.

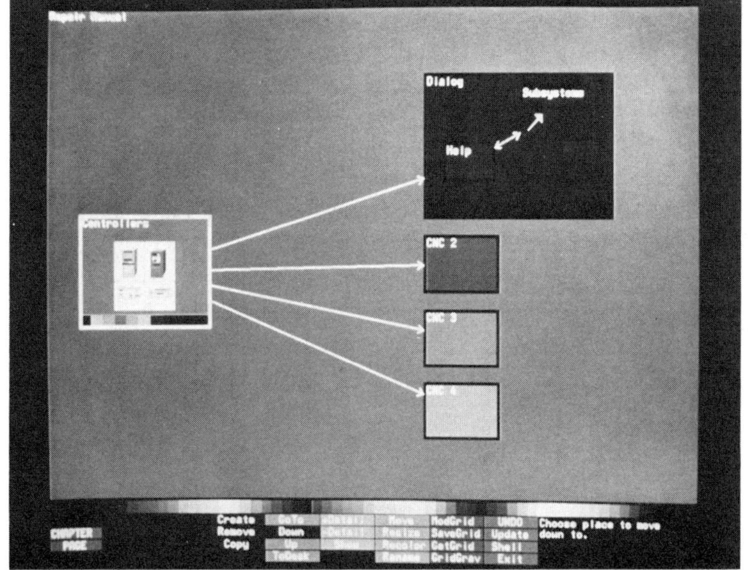

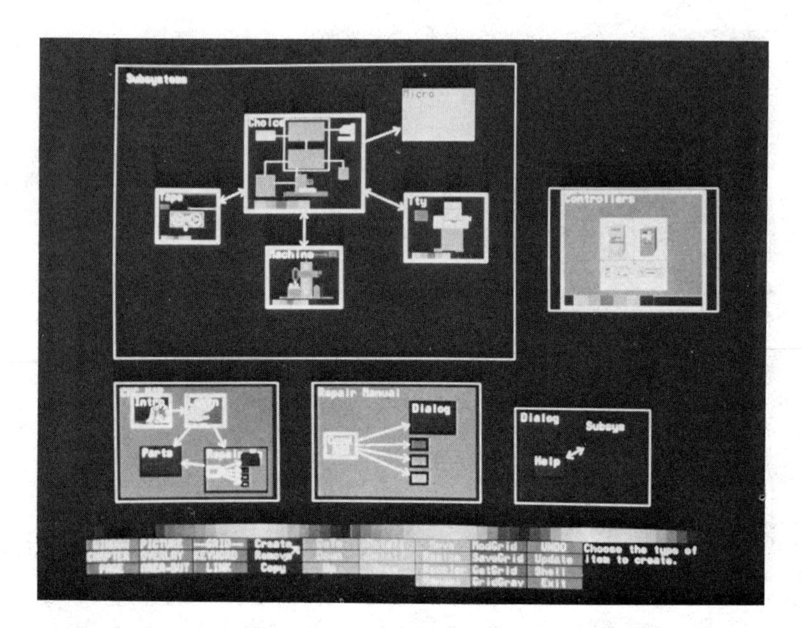

Figure six.

After pressing the "To Desk" button, the user may survey different parts of the document at the same time. Each window shows a page or chapter in the document. The bottom middle window shows the chapter previously being inspected. From this perspective, links between different pages and chapters may be made by the author, or the contents of each window may be individually altered.

presented to the reader with the Document Presentation System. Here we see one page (figure seven) from a repair manual much like the one that our author was just making. The buttons along the bottom, created with the Document Layout System, allow the reader to have access to other pages in the document. Some of these are special pages that are created on the fly and are designed to show the reader where she or he is in the document.

The reader of a conventional book can quickly look back at recently read pages. The designer of this document provided a similar facility for its reader by providing a "Timeline" button that creates a page containing miniatures of the most recently viewed pages in bands representing their chapters. (Figure eight.)

Figure seven.

A single page from a repair manual being viewed with the document presentation system. The buttons on the screen allow the user to move to other parts of the document.

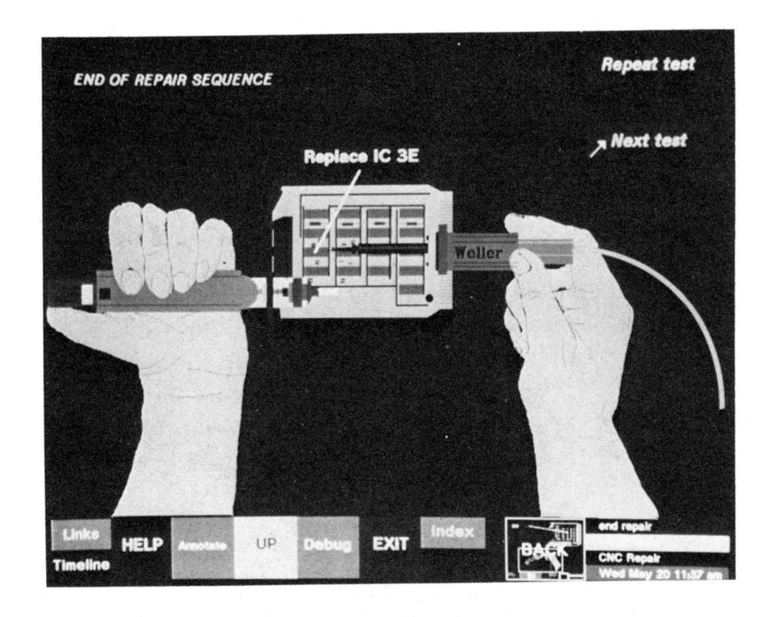

Figure eight.

The "Timeline" page showing reduced images of the pages viewed, in the order in which the user looked at them. The miniature images are shown in colored bands that represent their parent chapters. Touching the miniature page prompts it to grow to fill the screen.

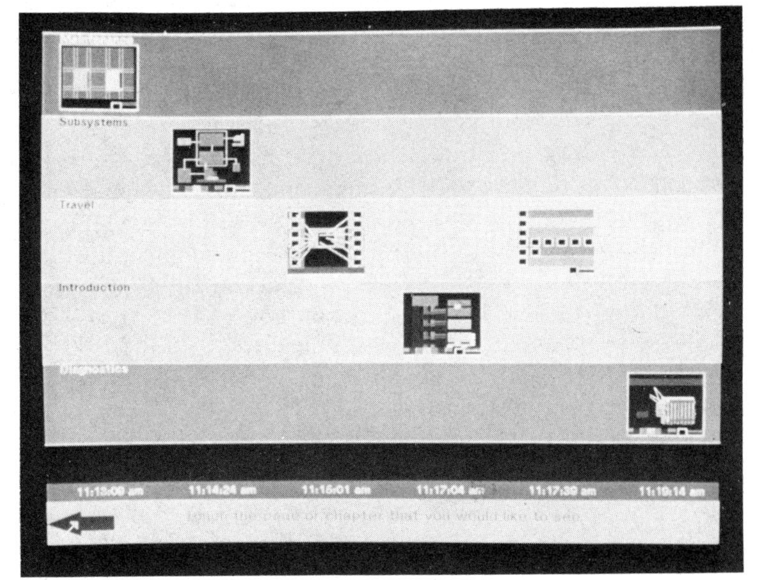

The "Links" button creates a page that shows a miniature of the current page in its chapter, surrounded by all of its possible predecessor and successor pages. (Figure nine.) Arrows at the bottom let the reader scroll the miniatures if there are more than can be comfortably displayed on the screen.

The "Index" button dims the screen and overlays a list of key words and phrases which the reader can pick. Miniatures of those pages that are associated with the selected word or phrase are then displayed (figure ten).

The "Timeline", "Links", and "Index" pages provide the ability to move around in the document, as well as showing its structure. Each page miniature that they display is a button that if touched will take the reader to the page that it represents.

One good question to ask oneself continually when working on a project such as this is, "What is wrong with what has been done so far and how can we improve it?" Looking at what we have accomplished, a number of very serious problems are evident. One of them is that creating a document with our system is extremely labor intensive. Most of the documents that we have written so far are maintenance and repair manuals, whose pages typically contain only a few sentences and some accompanying pictures. Given that, one might well imagine that writing a manual that explains how to maintain and repair a very large piece of equipment will take not just hundreds of pages, not thousands of pages, but tens of thousands of pages. Each one of these, and the even larger number of links that interconnect them, would be created by a combination of artists, designers, authors, and assistants. This is a time consuming process that can take many years of effort.

Much of this work is made necessary by the passivity of the authoring system that we are using. It is a willing servant, there

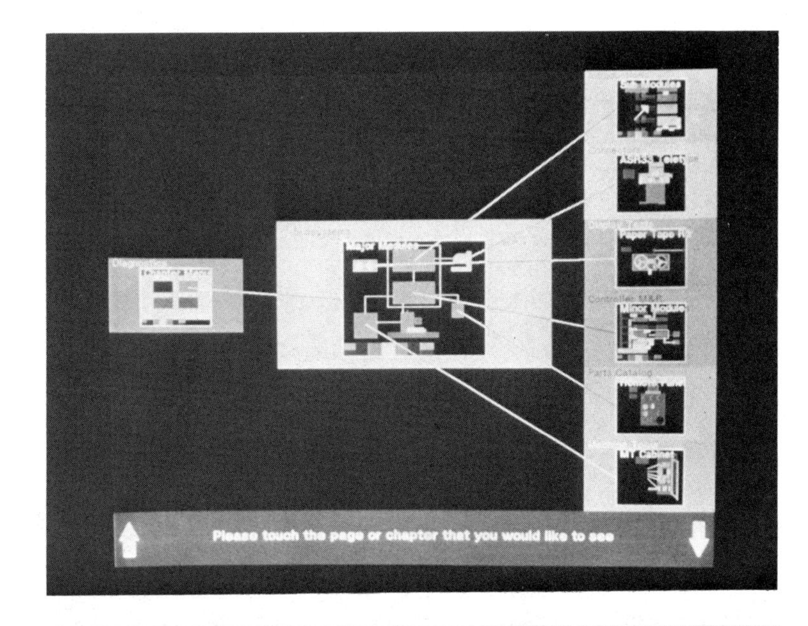

Figure nine.

The "Links" button will create a page like the one above. It shows, on the left, all the preceding pages, and, on the right, the succeeding pages. If the user wants to view any of these pages more closely she or he may touch the selected page and it will enlarge to full-screen size. The arrows at the bottom of the page are used to scroll forward or backward to see those pages that will not fit on the screen.

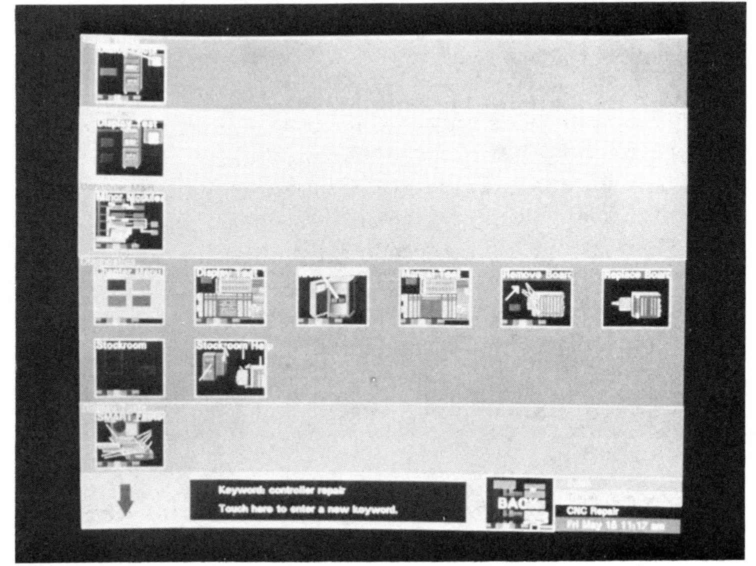

Figure ten.

The "Index" page displays pages related to a key word or phrase selected by the user. The colored bands in which the reduced pages appear, represent their parent chapters. If there are more page references than the screen will contain, the user may scroll ahead by touching the arrow at the bottom of the screen. Any one of the page references may be viewed full-screen if the user simply touches the selected pages references.

to do what it is told. If it is instructed to make a page it will oblige. It does not enforce any higher level of error checking than the syntactic. A chapter does not belong inside of a page so it will not let an author put one there. On the other hand, it will not say "These are the pages that you should consider putting in this chapter." Tedium is not being eliminated in the process of document creation. Once a style has been established by a designer, for example, it can be plain drudgery for a person to follow the pattern over and over again in page after page.

The project we are now working on is called Automated Authoring, and it is our hope that it will help reduce the immense amount of work required by document creation. In Automated Authoring, we are trying to create a system that can follow a set of design rules to make pages by itself. The final product will be a series of pages, containing text, pictures, or even sound, that explains how to do something. Our emphasis is therefore on automating the creation of documents that are explanatory, like the maintenance and repair manuals we created with our previous system.

There are three major components in our Automated Authoring project. The first is the "Problem Solver". Its job is to take the high level tasks that the document is expected to instruct its reader how to perform and break them down into lower level tasks. We call the second component the "Communicator". It will determine how to explain each lower level task using picture and text and how to develop both the sequence of pages and the individual contents of each page. The third component is the "Production Assistant". It provides an interface to other software which sets text and generates needed pictures.

I'll go into a bit more detail about each of these systems. The Problem Solver is given a problem which it breaks into a number of sub-problems, each of which is small enough by itself to be easily solved. It uses a set of rules for solving problems in general, as well as rules designed for a particular problem domain. Suppose, for example, that our high level problem is to remove a circuit board from its cabinet. The Problem Solver might break this down into finding the cabinet, turning off its power, opening the door, etc. Problem solvers of various sorts are a topic of research in artificial intelligence. We are using one that is currently being developed at Brown University.

The output of the Problem Solver is expressed in a low level semantic notation, comprehensible to a computer scientist, but hardly the sort of thing that we would give to someone who actually had to remove the circuit board. This is where the Communicator comes in. It has three major jobs. First, it must figure out on which page to explain each of the sub-tasks output by the Problem Solver. Next, it must turn each 'sub-task' on a page into the appropriate pictures and text that are needed to explain it. For example, it must come up with an explicit description of the contents of a picture and caption that explain to a person that they should "Turn the handle slowly." Finally, it must lay out a page, positioning each of the pictures and pieces of text.

The result of the Communicator's work is an explicit list of instructions for everything that will appear on each page. This list is fed to the Production Assistant, which interfaces with other programs that actually create the pictures (or retrieve them from a library) and set the type in place on the page. Some of our other projects at Brown are concerned with the actual generation of realistically shaded pictures of objects given information such as lighting and camera position. We will be using them to generate the pictures.

What I have just described is an ambitious ideal. We have a

Footnote

This work is supported in part by the Office of Naval Research under contract N00014-78-C-0396 Andries Van Dam, Principal Investigator.

simple "scratch" version of our Automated Authoring system running now which neatly sidesteps many very difficult problems that we encountered in its creation. In particular, this preliminary version discharges its task of creating pictures for each task by ordering a sort of group snapshot of all the objects discussed on a page without any consideration for how the objects are used. It also lays out its pages in a rather uninteresting way. Most of our current work is concentrated on these two problems. We are formalizing two sets of rules: one for creating explanatory pictures, another for laying out the pages that will contain them.

To sum up, we are trying to create a system that, by following rules of the sort employed by designers, is capable of producing a competently designed and structured presentation. Part of the rationale behind our project is the desire to produce such presentations where the costs in human effort and time are currently prohibitive. There is another goal, however. As the computing sytems that we use become more powerful, we will eventually be able to generate "pages" on demand, nearly instantaneously, in what is more a conversation than a document. To maximize our own effectiveness in this approaching era of abundant, cheap, and immediate information, it is essential that we be presented with clear, well designed graphics.

Programming by Rehearsal: A Graphic Programming Language for Computer-Based Curriculum Design

William Finzer

Computers as classroom learning tools are not living up to their promise. The presentation of information on these classroom terminal screens looks dismally similar to that found on textbook pages. Interactive systems and new computer programming languages may help realize the educational power of computers.

Designing curriculum materials that use computers for school classrooms is a vital task today for which there are too few tools available. The most interesting and potentially valuable educational uses of computers involve simulations and "exploratory environments" and yet these are the most difficult kinds of applications both to design and to implement. The alternative, which one too often sees, is the duplication of textbook pages on the screen along with multiple choice questions.

If we wish to see more creative uses of computers in education, we must ask "Who will be the people who will design and implement them?" Very few teachers or curriculum designers are also capable programmers, and indeed, one would prefer that they devote their own training to learning about how children learn, not to programming. The designer then, must work with a programmer. There are two unfortunate consequences of this fact.

1 The design of the curriculum materials is not interactive. The designer specifies, usually on paper, what he or she wants and then must wait several days, weeks, or months, to see what this looks like on the screen - a cycle which must be repeated even for minor changes.

2 Extensive experimentation does not happen when the programmer works alone to implement the design. After all, the idea belongs to the designer, not the programmer. When the designer and programmer are combined in a single person, one more often finds playful experimentation and serendipitous creation. The designer's imagination is free to roam over the possibilities for using the computer.

The implementation of new ideas is painfully slow even if the designer is also a programmer. Although today's high level programming languages represent a tremendous advance over machine language programming, there remains an incredible amount of detail which must be programmed to implement a reasonably complex computer simulation or learning environment.

Aims

At the Xerox Palo Alto Research Center (P.A.R.C.), both my colleague, Laura Gould, and I, have designed computer-using curricula over the past several years. Although we are both programmers, we have still felt the need for a design tool that would allow us to think more about designing and less about programming. And we have been particularly interested in putting such a tool in the hands of people who are not normally thought of as programmers.

I will describe here the results to date of our continuing work, which we call "Programming by Rehearsal." We implemented this system in Smalltalk-80, a programming language developed in 12 years of research and language design by the software concepts group at the Palo Alto Research Center.

Our aims are that teachers and curriculum designers (and students, for that matter) be able to produce computer programs of high quality, by which we mean that they would be exploratory environments, or "microworlds" (to use Seymour Papert's terminology). The users should be able to create these programs quickly and interactively, in experimentation mode, much as an artist might sketch out an idea on paper, gradually refining it to become the finished drawing. And, curriculum designers and teachers should be able to make computer programs by themselves without having either to employ a programmer, or to be one. (Although we believe in a team approach to curriculum design, we do not believe that it would be necessary for one of the team members to be a programmer.)

Properties of the Resulting System

The system we created has several important properties. Layout is done graphically, much as one would sketch on paper, without having to worry about screen coordinates. To a larger extent, what you see is what you get. There are some tools, like gridding, which aid in this layout process. The system is complex, but its capabilities are easily discovered. If the system is to be powerful enough and flexible enough to produce truly interesting programs without involving the designer in overwhelmingly detailed work, then it will also provide very high level, varied features. There must, however, be easy ways to discover what features are available and to try them out. In the Rehearsal world you can find out about system capabilities easily both by using a large and powerful help system, and by simply trying things out, that is, "rehearsing" them. There is no need to refer to a bulky, difficult-to-read manual.

Some Finished Examples

If you had never seen a house, I would not try to explain how to build one without first showing you what a house looks like. So, before describing how we make things using Programming by Rehearsal, I shall first illustrate some simple "productions" made using it. These particular productions were chosen not because they have significant educational content, but because they give the flavor of what is possible and are extremely easy to construct.

The first, shown in figure one, is called "Quilting" and was made by Joan Ross, a curriculum developer from the University of Michigan, in the course of a month-long visit to the Palo Alto Research Center during the summer of 1982. During her stay with us, Joan played quite extensively with the Rehearsal world, working primarily on her own, and created a variety of simple mathematical learning activities.

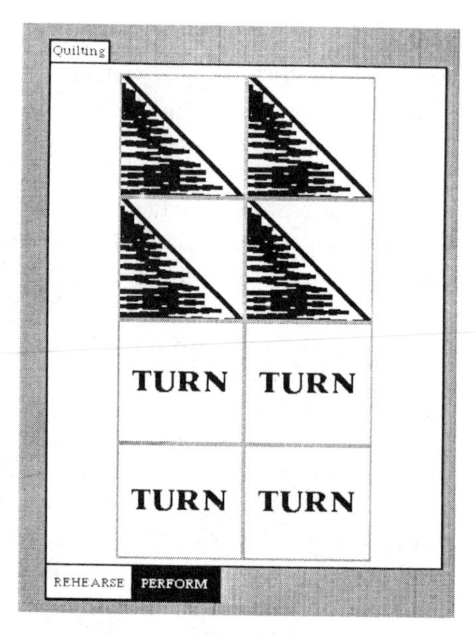

Figure one.
The Quilting example.

Figure two.
Four possible quilt pieces.

Figure three.
The Picture Play production.

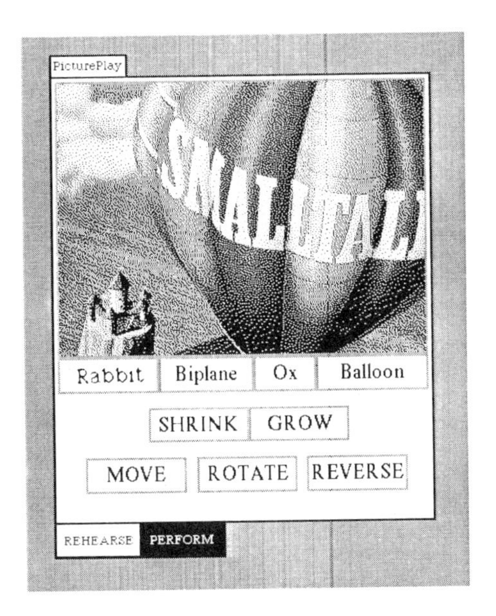

Each of the squares labeled "turn" is a "soft button". That is, the student can "push" it using a pointing device called a mouse. The mouse fits nicely under the student's hand and can be rolled around on the table. The mouse communicates with the computer through a cord (its tail). As the student moves the mouse on the table, an arrow-shaped cursor moves around to indicate positions on the screen. On the top of the mouse are three buttons colored red, yellow, and blue. The red button is used to point at or select things on the screen. In this example, selecting one of the turn buttons causes the corresponding quilt piece to turn by 90 degrees; 256 different patterns are possible, four more of which are shown below in figure two.

A second production is called "Picture Play" (figure three). It is driven by the soft buttons underneath the picture. These buttons allow the student to change pictures, to make the picture larger or smaller, to reverse the black and white in the picture, and to rotate or move the picture. Through play, the student comes to a beginning understanding of computer manipulation of bit-mapped pictures.

Each of these sample productions, while only entertaining in their present form, can be extended to encompass significant educational content. The point of emphasize here is that they can each be constructed in less than thirty minutes.

The Rehearsal Metaphor

Before I go on to consider exactly how productions like those shown above can be made in the programming by Rehearsal environment, I will introduce the basic metaphor. To misquote Shakespeare, "All the world's a theater". (A stage is just a part of the Rehearsal world.) In the rehearsal metaphor, we refer to the person who is putting together a new production as the "director" though that person also plays the role of playwright. The director looks through "Central Casting" (the list in the lower-right corner of figure four) to find appropriate "performers", "rehearses" them and writes their scripts.

The director can always ask for help by pressing the "help" button and will be given a message in the "prompter's box" explaining what the present possibilities are. (The long rectangular box at the bottom of figure four is the prompter's box.)

Every graphic object in the theater can be queried directly, by pointing at it with the yellow mouse button. The normal arrow cursor is then replaced by the name of that particular object: this name follows the motion of the mouse until it is where the user wants it to be. Then, the red mouse button is pressed to "drop" the name in a particular space. Dropping the name inside the help box produces a message explaining the capabilities of that object and how to interact with it. For example, selecting the "Production Library" in this manner

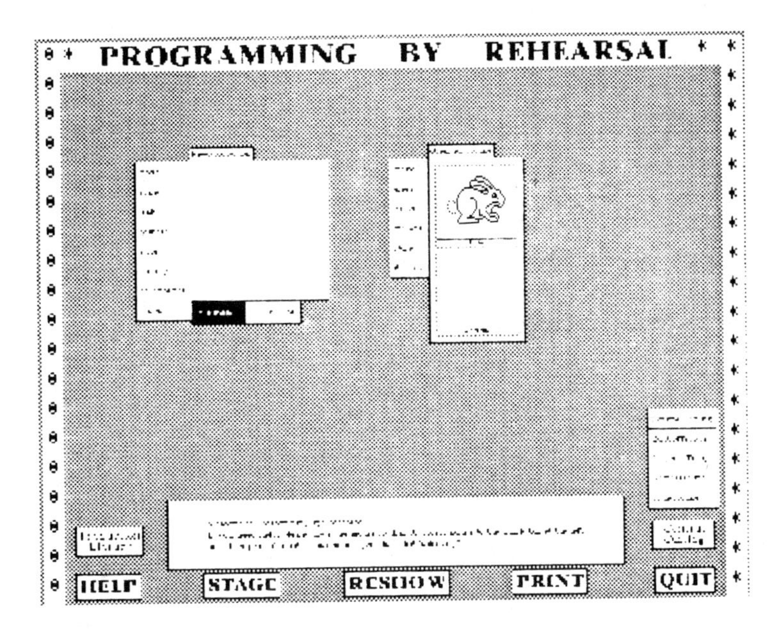

Figure four.
The whole Rehearsal world, showing an empty stage on the left and the Graphics Troupe with two performers in it on the right.

produces the following message.

"Production Library

If you select the button that reads Production Library, you will get a menu of stored productions. You can find out about a production by pointing at its menu name with the yellow mouse button. You can see what a production looks like by selecting its menu name with the red mouse button."

Performers are gathered together in "troupes". Figure five shows the "Graphics Troupe", which contains two performers, a picture and a turtle. (We will be dealing primarily with the picture. Turtles are modeled on the work of the MIT LOGO group and are basic tools for producing drawings. You can think of the triangle as representing a small animal which crawls around on the computer screen under your direction, leaving a line behind as a trace of wherever it has gone.)

A new production is put together by the director/playwright on a "stage". Figure four shows an empty stage labeled "New Production" to the left of the Graphics Troupe. Everything that appears on a stage or in a troupe is a performer. Some examples of performers are numbers, buttons, counters, pictures and turtles.

Each performer understands a set of "cues" which are grouped together into "categories". By selecting a performer with the mouse the director can see what these categories are (figure six).

By selecting one of the categories, like "image changing", the director can see the set of cues (figure seven). Lastly, by selecting a cue, the director "sends the cue" to the performer with the result that the performer shows the director how it responds. Figure seven shows the "reverse" cue selected and its effect on the performer. The remaining cues in the image changing categories all require parameters describing the amount or way in which the Picture should magnify, shrink, rotate, or reflect. These parameters may be expressed by typing into the box labeled "paste or type" before selecting the cue.

Production - Picture Play
To make a new performance, once we have an idea of what we want to see, we must find the performers that we believe will be needed, rehearse them to find their capabilities, lay them out on a stage as we wish, and write their scripts so that they will

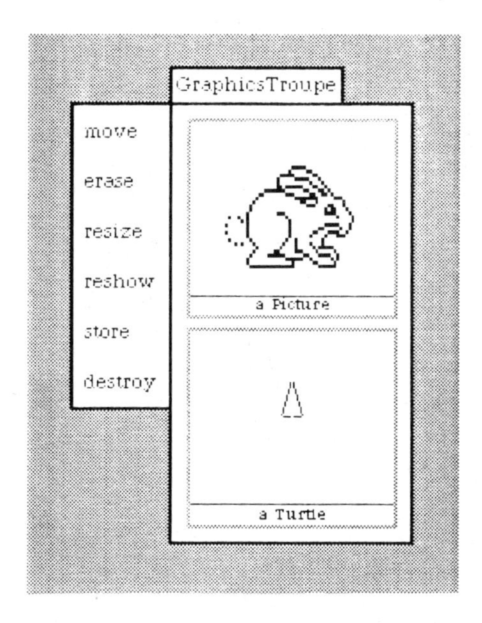

Figure five.
The Graphics Troupe.

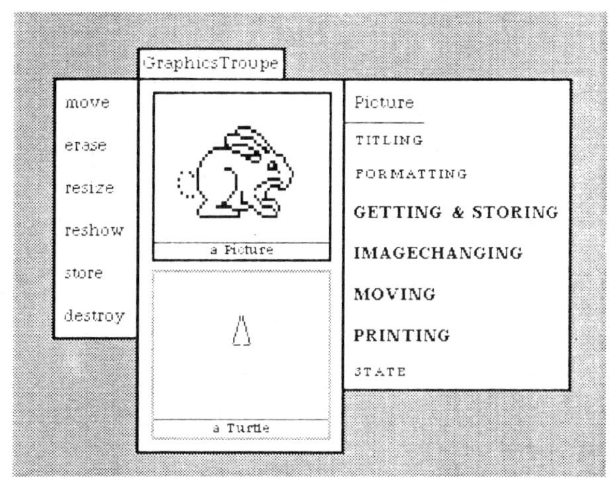

Figure six.
A picture has been selected and the categories of cues which it understands have been displayed along the right side of the troupe.

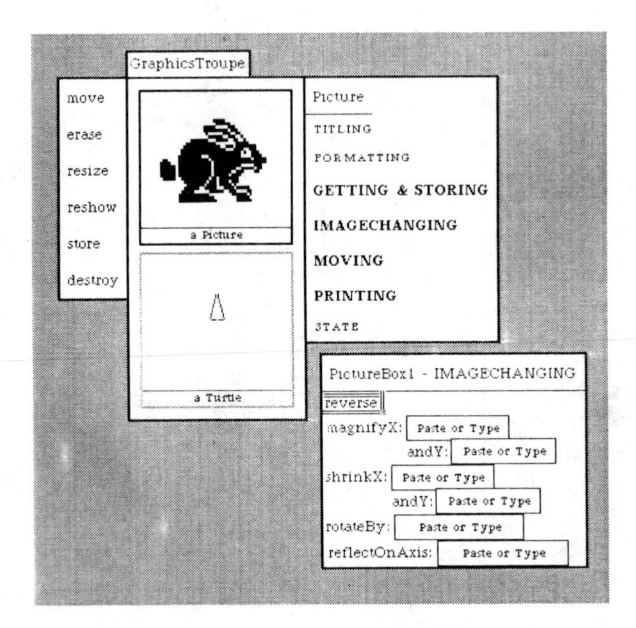

Figure seven.
The "image changing" category for a picture that has been selected, its cues displayed and "reverse"selected.

respond appropriately to the student user's commands. We expect that while this goes on, we will get new ideas about what the finished production should be.

Suppose that our goal is to produce the Picture Play example shown in figure three. There are only two kinds of performers in that production, a picture and nine buttons. We have already seen that pictures can be found in the Graphics Troupe. Buttons can be found in the Kernel Troupe, a troupe in which some generally useful performers are kept. (Figure eight.)

Every performer responds to a "copy" cue which we can use to pick up a copy of a picture from the Graphics Troupe, and to place it on an empty stage where we will make our new production. Similarly, we copy a button from the Kernel Troupe. (Figure nine.)

Figure eight.
Picture of Kernel Troupe.

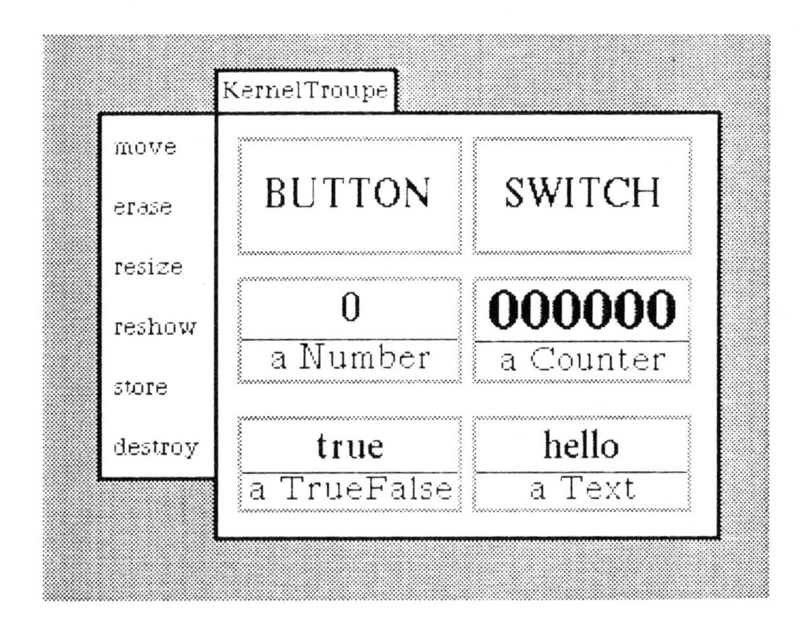

Figure nine.
Stage with picture and button.

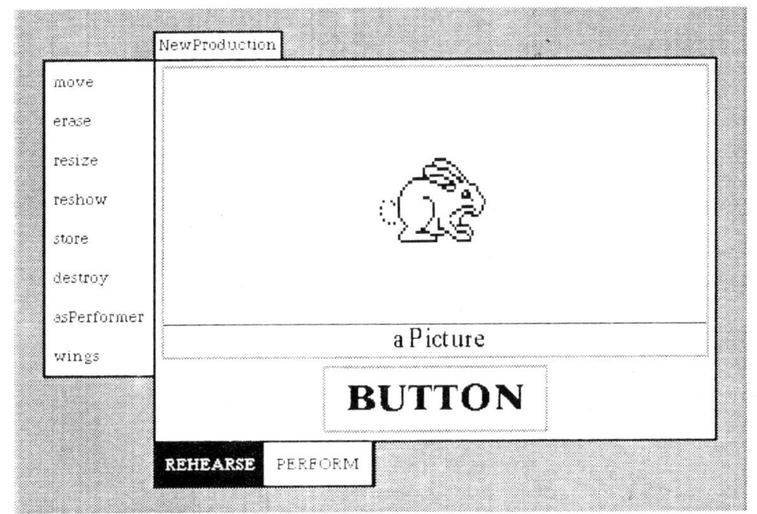

First we will change one of the buttons and, by doing so, will cause a particular picture, the biplane, to be displayed. The button's label should read "Biplane" instead of "button." Buttons respond to a cue, "relabel", which allows us to change their labels to whatever we wish. We call this "configuring" a button. (Figure ten.)

The special thing about buttons is that they have "action." We select a button's code "for on action" cue and get a "code box" which we will use for writing a code expressing what the button will do when the user presses it. The simplest method for doing this is to ask the code box to "watch" while we show it what we want to have happen. In this case we want the picture performer to display the biplane. (Figure 11.)

As we select the display "picture named" biplane cue, the picture changes appropriately and a line of code appears in the code box as shown in figure 12.

The buttons which the user will press to display the other pictures can be configured in exactly the same way. One example is show in figure 13.

Configuring a button to perform some operation on a picture is very similar to configuring one to display a new picture. We simply tell the code box for the new button to watch while we perform that operation. Figure 14 shows the code box for the reverse button just after it has watched the biplane execute the reverse cue.

The remaining buttons of the Picture Play production (shrink, grow, rotate, and move) were configured in exactly the same way. That is, their respective code boxes were made to watch while the picture was manipulated by sending it some cue.

Figure ten.
Titling categories.

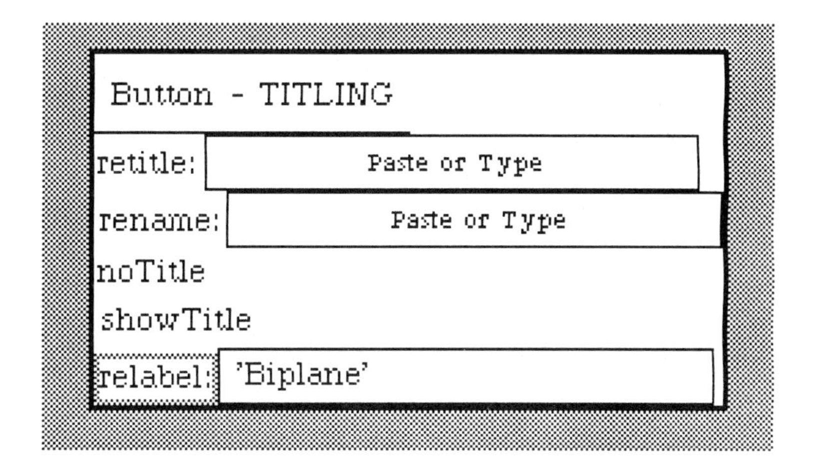

Figure 11.
A code box ready to watch.

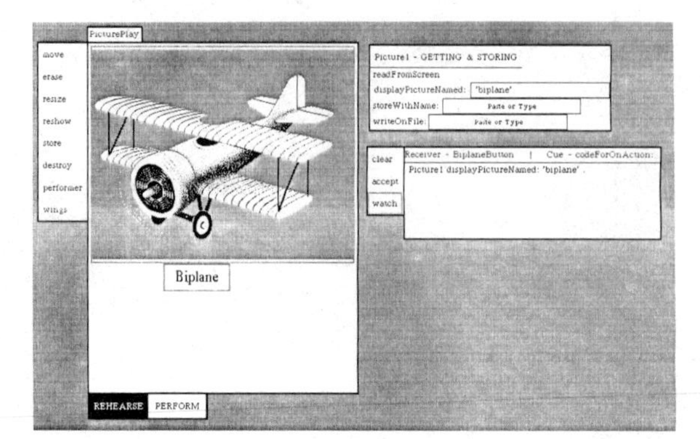

Figure 12.
After getting the biplane image.

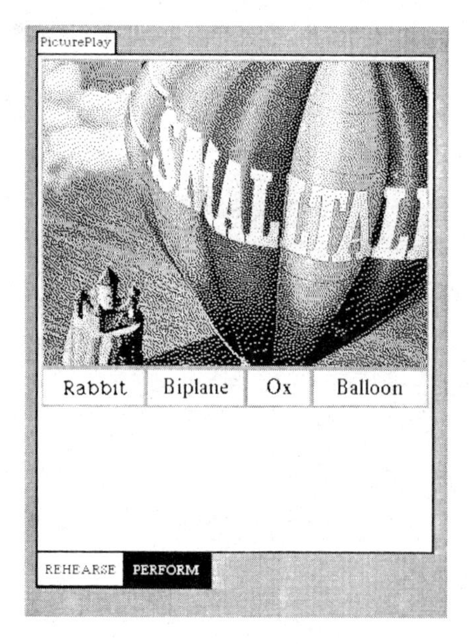

Figure 13.
The rest of the buttons for displaying pictures have been configured.

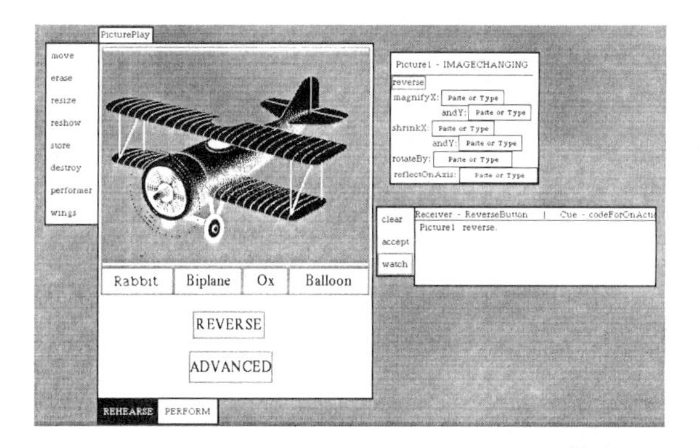

Figure 14.
After reversing the biplane. See below for a discussion of the "advanced" button.

Wings

In a real theater, not every performer is on stage at once - some are waiting in the wings for their entrances. The wings in Rehearsal serve several different kinds of functions.

In the Picture Play example we might wish to have a button labeled "advanced" which would provide additional buttons for manipulating the picture. These buttons are kept in the wings until needed. The action of the advanced button is to first erase itself and move the reverse button; then it brings the new buttons from the wings to the stage. (Figure 15.)

Sometimes we find it useful to leave a performer permanently in the wings. The "grow" and "shrink" buttons take parameters which specify the factor for growing or shrinking. While these factors can simply be numbers in the code which specify the actions of the two buttons, Rehearsal is made easier by placing

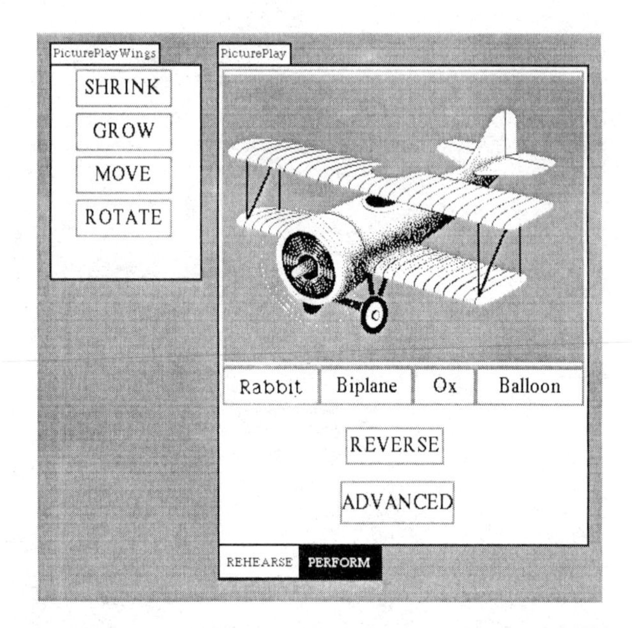

Figure 15.
The "advanced" button ready to call more buttons in from the wings.

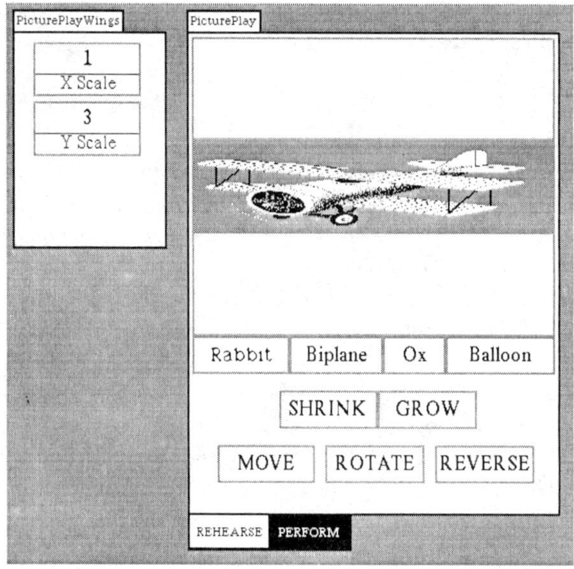

Figure 16.
"Grow" and "shrink" use parameters hidden in the wings.

two "number" performers in the wings and having the actions of grow and shrink refer to these new performers. (Figure 16.) As long as they stay in the wings, they will not be visible to the student, but will be accessible to us while we are designing. We can try different values for the grow and shrink factors: that is, "rehearse", until we find one that we like. (During this process, we might even decide that the student should also have this control over these factors, in which case we can move them from the wings onto the stage itself.)

The wings would be a good place to put a "coach " or "guide" performer. This kind of performer, especially useful in educational applications, would provide help for the student in the form of answers to questions, ideas about what to do next, and suggestions for better ways to do things.

Another permanent resident of the wings might be a "transcriber" performer. Its purpose would be to keep track of student actions and, on request, reproduce these actions either in the form of a written transcription or as a "movie".

Constructing and Modifying Performers

Any system that proposes to be flexible must itself be capable of modification. In the Rehearsal world, we find that we want both to be able to change the way a particular performer responds to a certain cue, and to add new cues. In addition, we want to create entirely new performers.

At present, changing cues or adding new ones cannot be done by the director in the Rehearsal world. In order to perform these functions, we have to drop through the "trap door" of the stage into the "Performer Factory" below. Here the sophisticated Smalltalk programmer can not only teach new cues to existing performers, but can create new performers as well.

At the Rehearsal world level, our directors can currently make new performers as long as they are constructed by aggregating old ones. This is done by selecting the stage's "performer" cue. For example, the Picture Play production can be made into a performer, and several of these performers placed on a single stage. The purpose of doing that might be to allow the student to compare the results of different graphic manipulations.

Future Directions

A major focus of our current research is on providing the capabilities of the Performer Factory at the Rehearsal world level. We will start by implementing a "cue maker" performer which will allow the director to define new cues for existing performers.

The system described here has been tested with just one real user so far. From that brief experience, we gained some confidence that the metaphor is an easily understood one and the user-interface provided is simple for naive users to interact with. We plan, over the next year, to invite other curriculum designers and teachers who are not themselves programmers to use the system. Then we can see what additional facilities we need to provide and how we can improve the user-interface.

The example programs we have constructed so far have not been very complex. We shall be exploring new kinds of performers; and we want to begin to discover what class of learning activites are most easily constructed inside the Rehearsal world, and what kinds of learning activities might not be constructible.

A criterion for success, which we keep in our minds as we further develop the Rehearsal world, is that not only teachers and curriculum designers, but students themselves should be able to use it for exploration and creation. We know that if the Rehearsal world is both accessible to children, and powerful enough for designers to create interesting educational applications, then it has the potential for wide use and therefore, significant impact.

Our thesis has been that teachers and curriculum designers who are not expert computer programmers can, and should, be able to explore ways to produce interactive, graphically oriented learning environments for students with the computer, and that the results of this exploration will be better than those obtained by using programmers to implement paper and pencil designs. The power of the dynamic medium provided by the computer should be made directly accessible to the designer, the teacher, and the student. If we, and others who are attacking the problem of accessibility, are successful, the result should be higher quality educational software than we have seen so far; more experimentation by designers and teachers with computer learning environments than has been possible before; and a richer experience for students as they use computers as tools for learning.

Computers in Design Education

Charles Owen

Significant work is being done at the Illinois Institute of Technology's Institute of Design to teach students about the computer technology they will face as professional designers.

Computers in Design Education

In a talk preceding the panel discussion, Professor Charles Owen described student projects and classes at the Illinois Institute of Technology's Institute of Design that make use of computers in some advanced and innovative ways.

At the Illinois Institute of Techology (IIT), all students are required to take a course in programming, whether they are English majors, engineering majors, or design majors. They begin with BASIC programming and, once they get past the simple problem solving level, move on to FORTRAN. This background gives design students a fundamental knowledge about computer systems, the procedures for programming them, and a foundation for design course work using computers.

Even though it is a popular application, the computer related courses at the Institute of Design (ID) are not all concerned with graphics - mainly because you can do much more with computers and design than graphics. Computers are a natural aid for designers, employing any kind of design method makes that clear. The elements brought to light through design method techniques can be enormous, sprawling, and seemingly unmanageable. The computer provides the support and organizational capacity that allows the designer to fit the elements together.

Computers at ID are used to, among other things, describe and evaluate design problems, they are used to structure complex informaton and show the relationships between those pieces of information. Students can, if they like, get heavily involved in computer work at ID.

The Facility

Students have access at IIT to a large Univac 1100/81 computer which is used with keypunch and terminals. (Figures one and two.) There is also a Prime computer system - a large interactive mini-computer accessed through terminals available to students across the campus. These systems let students get the experience of working with two kinds of programming.

In January 1982, the facilities were greatly enhanced with the opening of a new Design Processes Laboratory. With support from the Hewlett-Packard Corporation, the lab opened with a new computer with 120 million bytes of memory; a high quality micro-computer that serves as both a monitor and a graphics terminal; a graphic printer that can print 400 lines-per-minute; a small eight-pen plotter; a touch sensitive display terminal; and a digitizer which allows two-dimensional drawings to be translated numerically and stored by the computer. (Figures three through nine.) The new lab will be used for class and thesis work, as well as being used for departmental and supported research.

Figure one.

Computer room at the I.I.T. computing center. Four computers serve the campus, they are: Univac 1100/81, Prime 400, Prime 550, and DEC VAX 11/780.

Courses and Methods

Students in the Design Methods class learn about the basic design model by working through the steps of problem description and evaluation. In the process, specific tools, or methods, are employed, among them morphological analysis, synectics, and delphi. In many cases, the projects are completed with computer support, and use programs written by the students themselves.

Most design projects require using a great deal of information. Unfortunately, most designers use less information than they would like to, and less than they should, often because it is just too difficult to handle it all. Structuring the information allows more of it to be used, and to be used more easily. This is one of the things students learn in Design Methods.

Figure two

Data preparation area: rooms of keypunches and terminals serve students in several areas around the campus.

Although mathematics are an integral part of the course, it is a more descriptive and conceptual sort of math than that used for proving theorems, and constructing equations. Graphs are used as models for demonstrating networks of information. For structuring information in the computer, we use three computer programs: RELATN, VTCON, and CHART. The organizing concepts used and discussed in class are ones essential and

Figure three.

The Institute of Design's new Hewlett-Packard 1000F computer, disc memory and some peripheral devices. The large cabinets hold the computer, although what is in there is mostly air - the computer itself is less than one cubic foot. The rest of the space is for boards and a tape drive that allows communication with other equipment and other computers. On the far right is the fixed disc drive with 120 million bytes of memory.

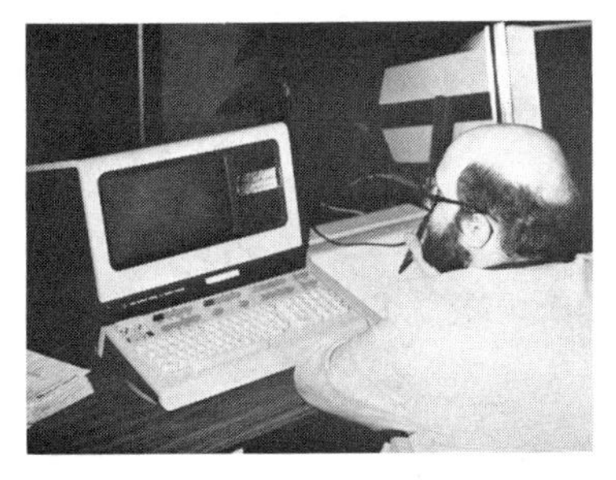

Figure four.

The Hewlett-Packard graphics terminal. This terminal is actually a very high quality micro-computer, serving here as a system monitor and a graphics terminal.

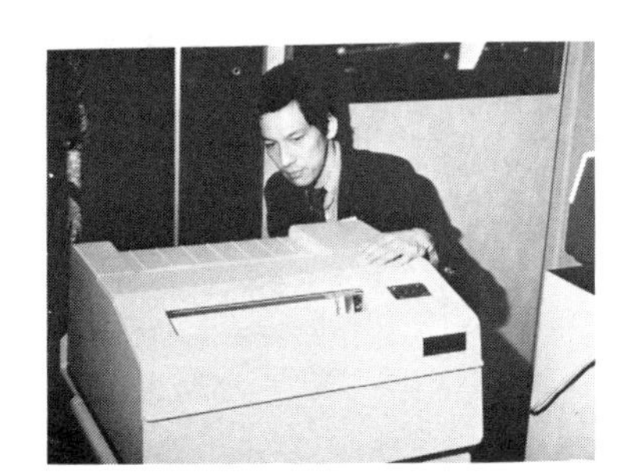

Figure five.

The Hewlett-Packard 400 line-per-minute graphic printer.

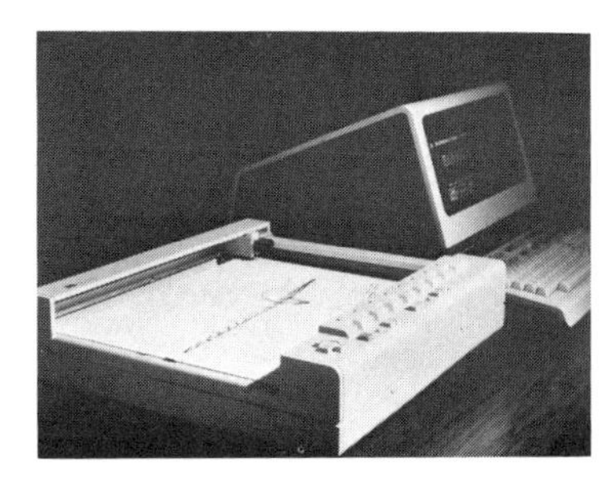

Figure six.

Hewlett-Packard's 11x14 inch eight-pen plotter, shown here with one student's three-dimensional project for the graphics class.

Figure seven.

A Magnavox plasma screen terminal with touch panel. The panel can be programmed to recognize finger touch on the screen. Holes around the side of the screen project infrared rays across the screen on a half inch grid. When the screen is touched, the grid is broken and the computer is able to locate the person's finger. The touch panel can be programmed to control rotations, transformations, create different forms and so on.

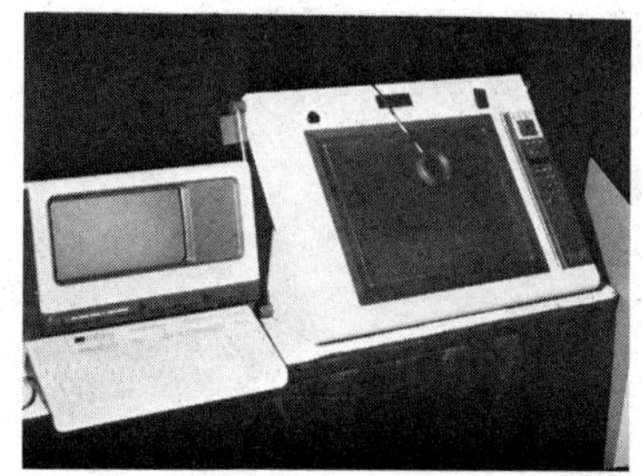

Figure eight.

Hewlett-Packard's high resolution digitizer. This allows two-dimensional drawings to be digitized by using a cursor to locate points and enter them as X and Y coordinates. It is a very good digitizer; 35 miles can actually be digitized in X and Y by locating reference points on the paper and re-establishing those points when the paper is moved.

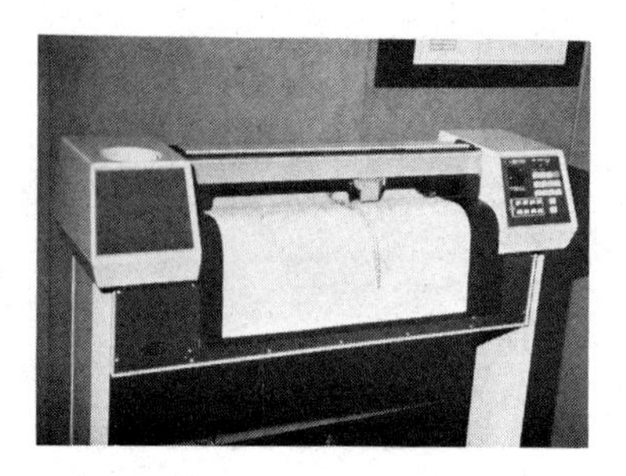

Figure nine.

The Hewlett-Packard large format eight-pen plotter. This plotter has a range of marvelous features: it will take paper up to 26 x 48 inches; individual sheets of paper can be used instead of a roll; it picks the paper up through the front, draws it in, finds the front edge, and then adjusts it; it plots to 1/1,000 of an inch accuracy; the paper can be taken out, put back, and the plotter will return to within 2/1,000 of an inch from where it left off, and continue on.

common to patterns employed every day in everyone's thinking habits, hierarchy, order, relationships, similarity and interaction. These are powerful and abstract concepts; in this class an attempt is made to formalize them.

One student used the computer to help with a project evaluating the effectiveness of the logos of selected visual design offices in Chicago. In approaching the project, the student asked,"How is the image of the company perceived by potential graphic design clients?"

Potential clients were given a list of antonyms like sharp/dull, or active/passive, with numeric levels of gradation in between the word opposites to indicate degrees of response. The collected responses were fed into the computer, which then positioned the reactions on a semantic differential map. (Figure ten.) One respected firm fared rather badly. From its position on the map, it was apparent that the RVI logo was not perceived positively by the people tested, people who were considering using a design company's services. One person who took the semantic differential test commented on the RVI logo, saying, "I thought they treated their identity too casually - almost offhandedly. Maybe they would treat my business the same way." Another person said the RVI logo was "too passive."

There are also ways of organizing the information used in design processes so that students can work in teams. They work with clients and users, finding insights into a behavior and why it occurs, and they record those insights so they will not be lost or forgotten. Once described, these become information elements to be entered into the system so the computer can determine which elements are related to one another. A matrix showing all the relationships can be produced by the computer, but seldom is because such a diagram would be so large and complex as to be of little value. Instead, a computer produced cluster diagram is made using the VTCON program. (Figure 11.) The cluster will show the relation of information elements as they pertain to one specifically prescribed problem. The resulting information structure is a road map to find the way through the design problem to a solution.

In the Systems Design course, students draw on all the resources they have developed in Design Methods by applying

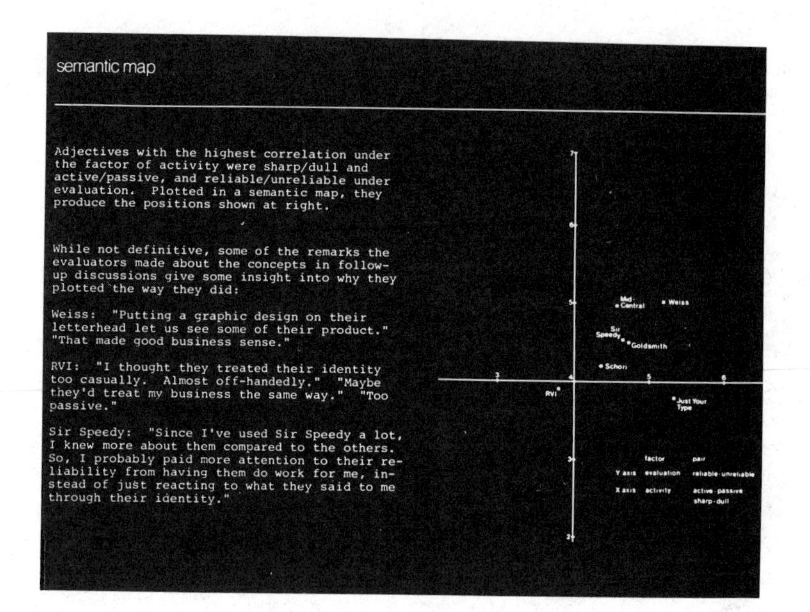

Figure ten.

Page from a report employing computer supported semantic differential techniques to evaluate the impact of letterhead designs used by visual design offices in Chicago.

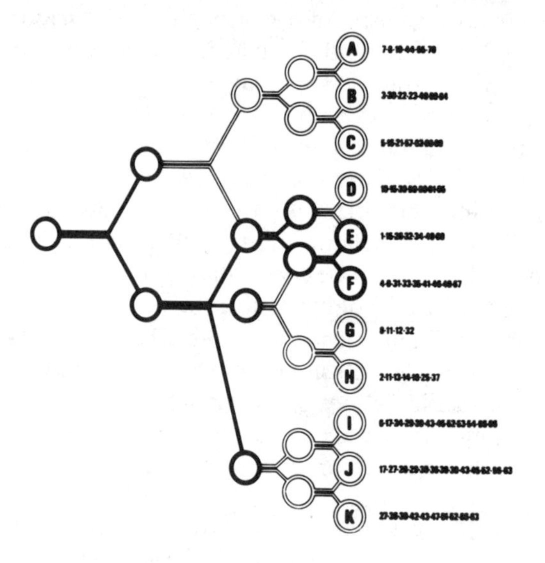

Figure 11.

An information structure developed by the VTCON computer program showing the design problem represented on the left, and the related information elements on the right.

structured planning concepts to very large and complicated projects. Working in teams, students generally take their project through to a conceptual solution. Often, a series of models is used to demonstrate the relationships between all the different elements of the design problem and its solution. In the course, students have attacked such difficult problems as: designing a major transportation hub that must accomodate a variety of transportation modes like trains, cars, and planes (figure 12); or creating a playground design that would allow both handicapped children and children of normal capabilities to play.

The design solution for the playground problem is very inventive; all the structures created for it are inflatable and erected by the children using air "wands" plugged into a central air bank. (Figure 13.) Children can build and structure this playground any way they like. The "Playscape" design won third place at the International Rehabilitation Engineering conference in 1979.

Control Technology in Design is one of the more basic classes, and the student's introduction to micro-computers. In this course, students become involved in actually building micro-computers. (Figure 14.) The purpose behind the course is to familiarize students with the mechanics of a computer, so that when the student becomes a designer he or she will be able to work directly with the engineers and technicians in developing micro-computer controlled products. These skills will be very important for anyone interested in product design.

In the class, students are taught a smattering of electronics (all design students take two semesters of physics) and soldering, along with digital logic, and machine language programming. Students work with ET 3400 Heath kits. These are bread-boarding machines that can be used to design products.

Figure 12.

Diagrammatic model for a transportation center designed in the Systems Design class using computer supported design techniques.

Figure 13.

"Playscape" inflatable recreational environment designed in the Systems Design class. The large cylinder in the rear is the central air bank used to inflate the other units. The playground is designed to accomodate children with and without handicaps. This project won third place at the International Rehabilitation Engineering conference in 1979.

Figure 14.

Construction of a Heath ET 3400 micro-computer bread-boarding kit for work in the Control Technology class.

Students in one of the Control Technology classes constructed a mock film processor. (Figure 15.) The computer control was built and programmed by students to simulate the heating and cooling process in developing film. Their micro-processor is capable of controlling temperature within one-half degree, and will display temperature readings on its screen. Another micro-processor controlled the sequence and timing of the processing.

What would qualify as ID's computer graphics course is called Computer Techniques for Image-Making. (There is already a computer graphics course in the computer science program, so we had to call ours something different.) After a semester's work, students in this class are working at sophisticated programming. The rudimentary skills learned in basic programming classes are developed in Computer Techniques for Image-Making so that students can learn to program with graphics.

Figure 15.

A mock-up of a film processor with major control processes simulated on ET 3400's.

They begin by working with simple X,Y coordinate systems. In time, they move to two-dimensional form transformations, using existing programs. After perhaps a two or three week project, they start to create their own three-dimensional program that will build three-dimensional shapes, will perform rotations, projections, and other complex operations. (Figures 16 and 17.) By the end of the twelfth week, everyone has his or her own operating program written in standard FORTRAN.

From this point, students move on to explore the differences between batch operations - using keypunch and big computers - and interactive operations where the effect of what the user is doing is immediate, and usually seen on a screen before him.

In their third project, students are working with advanced concepts. One example is a two-dimensional student project done by Constance White. (Figure 18.) Using the program called GROWTH, she began with a form similar to the letter y

Figure 16.

The Wright Brothers airplane as digitized and drawn by a program written by Kuohsiang Chen in the Computer Graphics class. This computer image contains about 2,500 points, and 3,000 lines. It could be "rotated" and "flown".

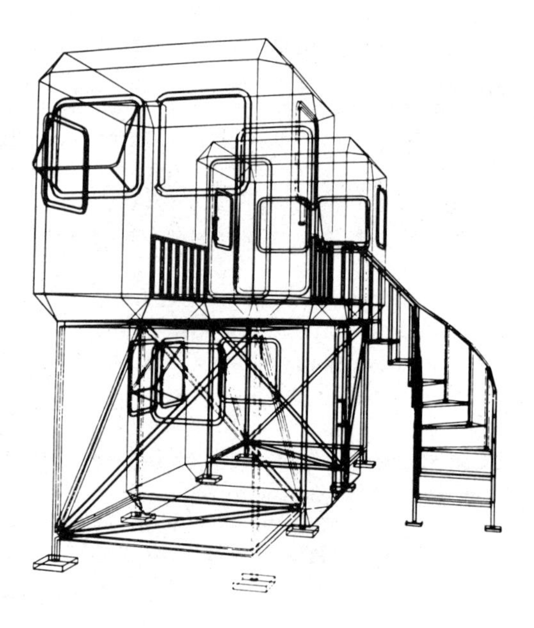

Figure 17.

A prefabricated housing system drawn in Computer Graphics class with a program written by Soon-Jong Lee.

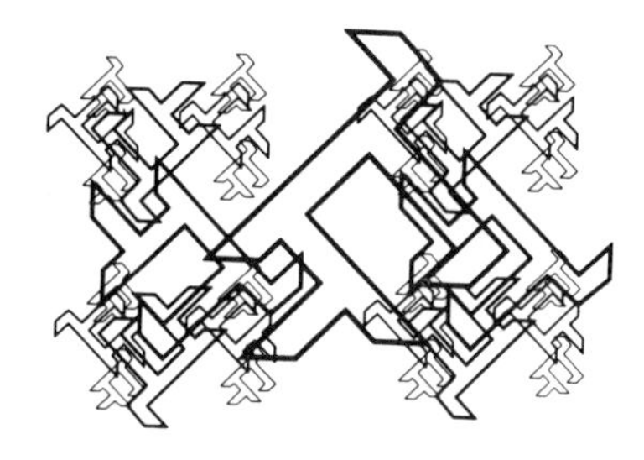

Figure 18.

A computer graphic pattern produced by the GROWTH program as a two-dimensional transformation project in the Techniques for Image-Making class.

which was used as a starting "seed". The form is treated like an organism. The user locates "bud" points and determines what the bud's relative size will be. The image grows over five "seasons" with new images budding at each specified location. In successive seasons, new buds may grow on the old, and so on. The health of the organism can be programmed to be less than perfect; some buds will not grow, others will change so as to be out of line, or different from the others in form. Interesting, totally different kinds of graphic images emerge this way. They are not the kinds of forms one would normally find in graphic design. Yet, this is using the computer the way it ought to be used - to generate new kinds of images which can only reasonably be done by computers.

Master's Projects

At the master's thesis level, students work extensively with computers, either using them for support, or in developing new computer processes. Thesis types have been divided into three basic groups at ID: contemporary problems, for which we look for generic solutions; design processes; and the most difficult, design theory.

In an interesting twist, one visual design graduate student, Eileen Gordy, wrote a program for her thesis project that shows how the computer can give back to typography what mass production had taken away. With her program, each letter can be individually spaced and drawn to accomodate and balance the letters to either side. (Figures 19 and 20.) With this careful design manipulation, it is possible to achieve, once again, the fine quality of the illuminated manuscripts. Solid type imposed the restriction of uniform spacing and form. But, with computer typesetting, letters are no longer physical representations of the letter forms; they are instead numeric representations that can be modified. The program that Eileen Gordy wrote instructs the computer to look at each letter -what is ahead and behind - and then change the letter forms so that

that each will balance. This allows subtle, individual adjustments, so that "TA", for example, can be treated differently than "RA". The T can be shortened or lengthened to fit better with the A, or the A can be modified to accomodate the R.

Figure 19.

Helvetica type conventionally spaced.

Figure 20.

Helvetica type computerized and modified in the typesetting process to automatically adjust letters toward optimum letter spacing.

Figure 21.

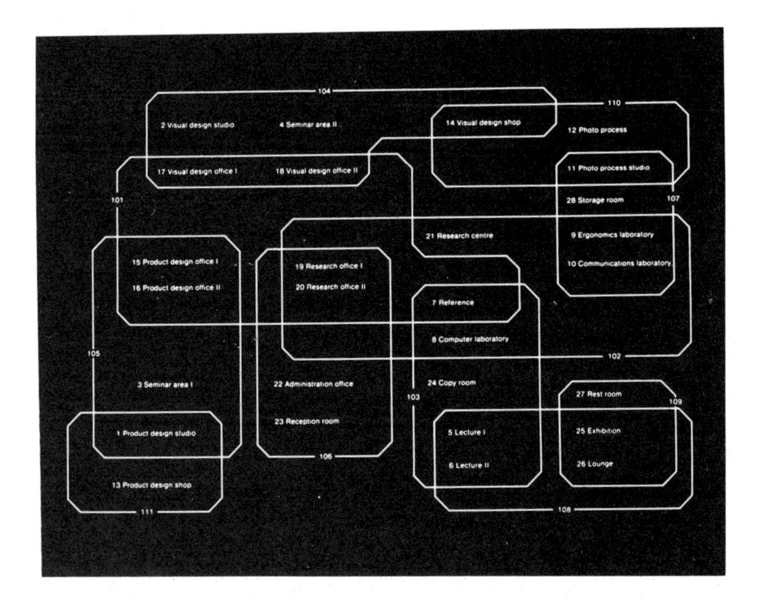

Another recently completed master's thesis dealt with optimization of arrangement. A new program made the computer capable of accepting a number of different criteria. Its task is to arrange, to the best possible advantage, the spaces in a building, for example, the parts in a product, or the events in a schedule. (Figure 21.) It is the best program we know of for this activity. A paper describing it in more detail was written by Keiichi Sato, and was published in the spring 1981 issue of Design Studies in England. In 1980, the same paper won the highest award at the Design Automation conference in the United States.

This gives some notion of what the program at the Institute of Design is like. It is changing already though, and moving forward. We plan to offer new courses in system theory and computer graphics application, a new graduate program is also being planned - all of which we find very exciting.

Biographies and Reading List

Jay Doblin has been a major force in both design practice and education. He was an executive designer with Raymond Loewy Associates, co-founder of Unimark International, and consultant to such major corporations as J.C. Penney, Standard Oil (Indiana), Gillette, and General Electric. He has served on a number of national and international commissions and delegations. From 1955 to 1969, he was the director and a professor at the Institute of Design in the Illinois Institute of Technology. In 1980, he was a fellow of the Royal Society of Arts in England, and a visiting professor at the London Business School where he conducted research and lectured about design. In recent years, Mr. Doblin has been working toward the development of a theory of design. Currently, he is the president of Jay Doblin and Associates where he consults to major corporations such as Xerox and Borg-Warner. At the same time, Mr. Doblin continues to write about design theory and teach at the Institute of Design.

Steven Feiner is a Ph.D. candidate in computer science at Brown University in Providence, Rhode Island. Mr. Feiner's interests span a wide variety of issues in computer graphics and human-machine interfaces, with an emphasis on computer animation and rule-based page layout. He is a member of A.C.M. SIGGRAPH and the Institute of Electrical and Electronic Engineers. Before entering the field of computer science, Mr. Feiner received a BA in music from Brown University.

Feiner, S., Nagy, S., and van Dam, A. "An Experimental System for Creating and Presenting Interactive Graphical Documents." **ACM Transactions on Graphics** (January 1982) 1:1, pp. 59-77.

Foley, J., and van Dam, A. **Fundamentals of Interactive Computer Graphics.** Reading: Addison-Wesley Co., Inc., 1982.

Nelson, T. H. **Computer Lib/Dream Machines**. Chicago: Hugo's Book Service, 1974.

Newman, W. and Sproull, R. **Principles of Interactive Computer Graphics**. New York: McGraw-Hill, 1979.

William Finzer is a Mina Shaughnessy Scholar in the mathematics department of San Francisco State University and a consultant in the software concepts Group of Xerox's Palo Alto Research Center. At San Francisco State, he has developed a computer-based mathematics curriculum for middle schools, courses for teachers on ways to use computers in the classroom, and is currently working on college level mathematics curriculum materials called "Windows", which use computers in exploratory modes. The work described here has been undertaken at the Palo Alto Research Center with Laura Gould.

Abelson, Harold, and diSessa, Andrea. **Turtle Geometry: The Computer as a Medium for Exploring Mathematics**. Cambridge: MIT Press, 1981.

Brown, Dean and Lewis, Joan. "The Process of Conceptualization." Educational Policy Research Center, Research Note EPRC-6747, December, 1968.

Goldberg, Adele, and Robson, David. **A Metaphor For User Interface Design**, proceedings of the Twelfth Hawaii International Conference on Systems Science. Honolulu: University of Hawaii, 1979, pp. 148-157.

Gould, Laura, and Finzer, William. "A Study of TRIP: A Computer System for Animating Time-Rate-Distance Problems." **Computers in Education**, proceedings of the IFIP TC-3 Third World Conference on Computers in Education. Lausanne, July 1981, pp. 359-366. Reprinted, "International Journal of Man-Machine Studies", volume 17, 1982, pp. 109-126.

Oettinger, Anthony G., with the collaboration of Sema Marks. **Run, Computer, Run: The Mythology of Educational Innovation.** Cambridge: Harvard University Press, 1969.

Papert, Seymour. **Mindstorms: Children, Computers, and Powerful Ideas**. New York: Basic Books, 1980.

Trowbridge, David, and Bork, Alfred. "Computer Based Learning Modules for Early Adolescence." **Computers in Education**, proceedings of the IFIP TC-3 Third World Conference on Computers in Education. Lausanne, July 1981, pp. 325-328.

Kristina Hooper is currently a research scientist in corporate research at Atari, Inc. Before joining Atari, Dr. Hooper was an assistant professor of psychology at the University of California, Santa Cruz. She has a Ph.D. in cognitive psychology from the University of California, San Diego. She has just completed work as the principal investigator of a National Science Foundation project which explored the use of computer graphics in teaching college mathematics. Before that, Dr. Hooper worked with the Architecture Machine Group at the Massachusetts Institute of Technology (MIT) as a visiting faculty member. She has also pursued post-doctoral work in architecture, dealing with environmental simulation and computer-assisted architecture at the University of California, Berkeley.

Arnheim, Rudolph, **Visual Thinking**. Berkeley: University of California Press, 1969.

Gregory, R.L. **Intelligent Eye**. New York: McGraw-Hill, 1970.

Halprin, Lawerence. **R.S.V.P. Cycles, Creative Processes and the Human Environment**. New York: George Braziller, Inc., 1970.

James, William. **Principles of Psychology**, Volumes I and II. New York: Dover Publications, Inc., 1950. (Reprinted from the 1890 original.)

Lindsay, Peter and Norman, Donald. **Human Information Processing**. New York: Academic Press, 1972.

McKim, Robert. **Experiences in Visual Thinking**. Belmont: Brooks/Cole, 1980.

Eric A. Hulteen worked from 1977-1982 with the Architecture Machine Group at the Massachusetts Institute of Technology (MIT). There, he was involved in developing concepts and software to facilitate the communication between man and machine. His work for his M.S. degree explored new styles of interaction between man and computer, made possible by the simultaneous use of speech recognition, gesture recognition, speech synthesis, computer graphics and eye-tracking. Mr. Hulteen received his undergraduate degree in architectural design from MIT. He is currently working for Atari, Inc. in California.

The Architecture Machine Group is a laboratory at MIT doing sponsored research in man-machine interactions and videodisc applications. It is one of the facilities at the forefront of the emerging field known as "media technology", a field that lies in the conjunction of three formerly separate disciplines -publishing, broadcasting, and computer science.

The Inferface to Design is an overview of Mr. Hulteen's work with the Architecture Machine Group. In it, Mr. Hulteen describes systems that can "understand" communications of speech, gesture, and point of regard, and which can respond using speech synthesis, videodiscs, and computer graphics. He will describe how computers can interact with users intelligently, and in a way more familiar to people.

Aaron Marcus is a graphic designer with 16 years of experience in computer graphics and information design. He holds a B.A. in physics from Princeton University, and an MFA in graphic design from Yale University. Mr. Marcus has recently opened his own consulting office. Aaron Marcus and Associates, with an emphasis on designing visual information systems for computer-based media. Before this, Mr. Marcus was a staff scientist in Lawrence Berkeley Laboratory's computer science and mathematics department. He has lectured, taught, exhibited, and published internationally in the fields of computer graphics, graphic design, visible language, semiotics, and the visual arts.

This article originally appeared in the July, 1981 issue of Centerline, published by the now-closed Center for Design in San Francisco. It later appeared in the March/April issue of Industrial Design Magazine, and is reprinted with the permission of Design Publications, Inc. Copyright Design Publications Inc.

New York, N.Y.
Mar/Apr 1982

Albers, Joseph. **Interaction of Color.** New Haven: Yale University Press, 1971.

Arnheim, Rudolph. **Visual Thinking.** Berkeley: University of California Press, 1974.

Benson, William H. "An Application of Fuzzy Set Theory to Data Display." Lawrence Berkeley Laboratory Technical Memo LBL-11590, 1980.

Bertin, J. **Semiologie Graphique**. Paris: Gauthier-Villars, 1967.

Boeke, Kees. **Cosmic View: The Universe in 40 Jumps**. New York: The John Day Company, 1957.

Bornstein, Marc H. "The Influence of Visual Perception on Culture." **American Anthropologist** Vol. 77, 1975, pp. 774-797.

Carter, James. "The Cartographics Potential of Low Cost Color Computer Graphics Systems." **Proceedings Auto-Carta** IV, 1971, pp. 364-371.

D'Andrade, R. and Egan, M. "The Colors of Emotion." **American Ethnologist** 1:1, 1974, 4a-64.

Dreyfuss, Henry. **Symbol Sourcebook**. New York: McGraw-Hill, 1972.

Hartley, James. **Designing Instructional Text**. New York: Nichols Publishing, 1979.

Herdeg, Walter, ed. **Graphis Diagrams**. Zurich: Graphis Press, 1975.

Joblore, George H, and Greenberg, Donald. "Color Spaces for Computer Graphics." **Computer Graphics** 12:3, 1978, pp. 20-25.

Kelly, Kenneth L. and Judd, Deane B. **Color: Universal Language and Dictionary of Names**. National Bureau of Standards Special Publication 440, U.S. Department of Commerce, 1976.

MacDonald-Ross, Michael. "How Numbers Are Shown: A Review of Research on the Presentation of Quantitative Data in Texts." **Audio-Visual Communication Review** 25, 1977, pp. 359-409.

MacGregor, A.J. **Graphics Simplified**. Toronto: University of Toronto Press, 1979.

Maier, Manfred. **Basic Principles of Design**. New York: Van Nostrand–Reinhold, 1977.

Muller-Brockman, Joseph. **The Graphic Artist and His Design Problems**. New York: Hastings House, 1968.

__________ **History of Visual Communication**. New York: Hastings House, 1976

Robinson, Arthur H. and Petchenik, Barbara Bartz. **The Nature of Maps**. Chicago: University of Chicago Press, 1976.

Robinson, Arthur H., Sale, Randall, and Morrison, Joel. **Elements of Cartography**. New York: John Wiley & Sons, 1959 and 1978.

Schmid, Calvin F. and Schmid, Stanton. **Handbook of Graphic Presentation**. New York: John Wiley & Sons, 1979 (second edition).

Wertheimer, Max. "Laws of Organization in Perceptual Forms." Ellis, Willis D., ed. **A Source Book of Gestalt Psychology**. New York: Harcourt Brace Jovanovich, 1939.

Marcus Articles

Marcus, A. "Designing the Face of an Interface." **Computer Graphics and Applications** (January 1982) pp. 23-29.

__________ "Computer-Assisted Chart Making From the Graphic Designer's Perspective." Lawrence Berkeley Laboratory Technical Report LBL-1-239, April, 1980.

__________ "Routes, Loops, Transfers, and Dead-Ends." **Print Magazine** 32:2 (1978) pp. 17-22.

__________ "At the Edge of Meaning." **Visible Language** 11:2 (1977) pp 4-20.

__________ "An Introduction to the Visual Syntax of Concrete Poetry." **Visible Language** 8:4 (Autumn 1974) pp. 333-360.

__________ "Symbolic Constructions." **Typographische Monatsblaetter** (St. Gallen) 92:10 (October 1973) pp. 671-683.

__________ "Computer-Aided Design: An Exploration." **Penrose Annual** (London) 66 (1973) pp. 191-198.

__________ "New Signs Along the Highway." **Print Magazine** 26:3 (June/July 1972) pp. 62-67.

__________ "A Prototypical Computerized Page Design System." **Visible Language** 5:3 (Summer 1971) pp. 196-220.

__________ "The Designer, the Computer, and Two-Way Communication." **Print Magazine** 25:4 (July/Aguust 1971) pp. 34-39.

Charles L. Mauro is president of C.L. Mauro Associates in New York City - a firm specializing in human factors engineering and industrial design. Mr. Mauro holds degrees in ergonomics and industrial design. He has received the Alexander C. William award, the Human Factors Society's highest honor for design, "for outstanding contributions to the design of a major operational man-machine system." Mr. Mauro has worked as the program manager for Henry Dreyfus Associates and for Raymond Loewy International. His articles have been published in a number of popular and professional magazines.

Ayoub, M.A. "The Problem of Occupational Safety." **Industrial Engineering** 7 (April 1975) p. 4.

Ducharme, R.E. "Women Workers Rate 'Male' Tools Inadequate." **Human Factors Society Bulletin** 20 (April 1977) p. 4.

Grandjean, E., Hunting, W., Sharer, R., and Wotzka, G. "An Ergonomic Investigation of Multipurpose Chairs." **Human Factors** 15 (1973) p. 3.

McCormick, E.J. **Human Factors in Engineering and Design.** New York: McGraw-Hill, 1982.

Nemecak, J. and Grandjean, E. "Results of an Ergonomics Investigation of Large Space Offices." **Human Factors** 15 (1973) p. 2.

Pearson, R.G. and Ayoub, M.A. "Ergonomics Aids Industrial Accidents and Injury Control." **Industrial Engineering** 7 (June 1975) p. 6.

Stellman, J.M. and Daum, S.M. **Work is Dangerous to Your Health**. New York: Vintage Books, 1973.

Tichauer, E.R. **Biochemical Basis of Ergonomics** New York: Wiley Interscience, 1978.

Wools, R. and Canter, D. "The Effect of the Meaning of Buildings on Behavior." **Applied Ergonomics** 1 (1970) p. 3.

Yoder, T.A., Lucas, R.L. and Botzum, G.D. "The Marriage of Human Factors and Safety in Industry." **Human Factors** 15 (June 1973) p. 3.

Reprints of most of the following are available upon written request addressed to:

Mauro Associates, Inc.
8 West 40th Street
New York, New York 10018

Bowen, Hugh M. "Human Factors Engineering." Maynard, Harold B., ed. **Handbook of Business Administration.** New York: McGraw-Hill, 1967 and 1982.

__________ "Product Liability: The Case of the Ten Cent Ring Guard." Presented at the Human Factors and Industrial Design in Consumer Products Symposium. Tufts University, Medford, Mass., May 1980.

__________ "Rational Design." **Industrial Design** i-vii (1964).

__________ "The Imp in the System." **Ergonomics** (1967).

Coskuntuna, Semra. "Instruction Manuals: Components of a Product's 'Teaching Package' ". Presented at the Human Factors and Industrial Design in Consumer Products Symposium. Tufts University, Medford, Mass., May 1980.

Mauro Associates, C.L. "The ManComputer Interface: A New Priority." **ManMachine 4** (Quarterly publication of C.L. Mauro Associates, Inc.) 1981.

__________ "Women in the Workplace: Equality and Equity." **WomanMachine 3** (Quarterly publication of C.L. Mauro Associates, Inc.) 1980.

__________ "Product Liability: For the Plaintiff and for the Defense." **ManMachine 2** (Quarterly publication of C.L. Mauro Associates, Inc.) 1980.

__________ "Three Mile Island: A Human Factors Problem." **ManMachine 1** (Quarterly publication of C.L. Mauro Associates, Inc.) 1979.

__________ "Human Factors Analysis of a Multi-Function Digital Watch." **Human Factors Society Annual Meeting Proceedings**, 1980.

Mauro, C.L. "From the Counter to the Courtroom." **Metropolis** (October 1981).

__________ "Human Factors Study Crucial for Future Office Computers." **Industrial Design** (March/April 1981).

__________ "Why Human Factors Engineering Will Pay Off in the 1980's - A Review of the Futurist Literature and Consumer Trends." Presented at the Human Factors and Industrial Design in Consumer Products Symposium. Tufts University, Medford, Mass., May 1980.

__________ "Three Mile Island: A Human Factors Problem." **Industrial Design** (September/October 1979).

__________ "How and Where to Find Research Literature." **Industrial Design** (March/April 1979).

__________ "Designing Against Product Liability Claims." **Industrial Design** (September/October 1978).

__________ "Can Hairdryers be Safer? Research Says 'Yes'." **Industrial Design** (May/June 1978).

__________. "How Human Variability Affects Design Part II." **Industrial Design** (January/February 1978).

__________. "Variability of the Physical Human is a Key to Design." **Industrial Design** (November/December 1977).

__________. "Abstract Models Improve the Product User Interface." **Industrial Design** (September/October 1977).

Mauro, C.L. and Bowen, Hugh M. "Why Human Factors Will Pay Off in the 1980's: A Review of Futurist Literature and Consumer Trends." **Human Factors Society Bulletin** (November 1980).

__________. "Properly Designed Offices Must Be Worker Support Systems." **Industrial Design** (November/December 1979).

__________. "Product Liability Claims Forcing Human Factors Emphasis." **Contract** (September 1978).

Charles Owen is professor of design for the Institute of Design (ID) at the Illinois Institute of Technology (IIT). There, Mr. Owen heads the product design programs for graduate and undergraduate students, and directs the Design Processes Laboratory. He is also a director of the Design Planning Group, Inc., and a consultant to Digital Equipment Corporation and S.P.S.S., Inc. (a computer software development company). He has worked in the fields of computer supported design, design methodology and computer graphics for more than fifteen years. He has written a number of computer programs for business and institutional applications. His current research is directed toward structured planning techniques, cellular solid modeling for computer-aided design and computer supported diagramming techniques.

Bowman, W.J. **Graphic Communication**. New York: John Wiley and Sons, 1968.

Chen, K. (in progress). **A Graphic System for Activity Analysis**. Chicago: Institute of Design, Illinois Institute of Technology.

David, R.A. **Proposal for a Diagrammatic Language for Design**. M.S. thesis. Chicago: Institute of Design, Illinois Institute of Technology, 1973.

Doblin, J. "The Map of Media." **Industrial Design Magazine**, January/.February 1981, pp. 35-37.

Garland, K. "Some General Characteristics Present in Diagrams Denoting Activity, Event and Relationship." **Information Design Journal**, 1979, pp. 15-22.

Herdeg, W., ed. **Graphis Diagrams**. Zurich: The Graphis Press, 1974.

ISSCO staff. **Choosing the Right Chart**. San Diego: Integrated Software Systems Corporation, 1981.

Karsten, K.G. **Charts and Graphs**. New York: Prentice Hall, Inc., 1923.

Murgio, M.P. **Communications Graphics**. New York: Van Nostrand-Reinhold Company, 1969.

Nelson, T.H. **Computer Lib and Dream Machines**. Chicago: Hugo's Book Service, 1974.

Nishimura, Y. (in progress). **Dynamic Information Display**, M.S. thesis. Chicago: Institute of Design, Illinois Institute of Technology.

American Standards Association. **Time Series Charts, A Manual of Design and Construction**. New York: Committee on Standards for Graphic Presentation, American Society of Mechanical Engineers, 1938.

Briton, W.C. **Graphic Methods for Presenting Facts**. New York: Arno Press, 1980. (Reprinted from The Engineering Magazine Company, New York, 1914.)

Edwards, J.A., Twyman, M. **Graphic Communication Through Isotope**. Reading: Department of Typography and Graphic Communication, University of Reading, 1975.

Lockwood, A. **Diagrams**. New York: Watson-Guptill, 1969.

Lutz, R.R. **Graphic Presentation Simplified**. New York: Funk & Wagnalls Company, 1949.

Neurath, O. **International Picture Language**. Reading: Department of Typography and Graphic Communication, University of Reading, 1980. (Facsimile reprint from 1936 edition.)

Schmid, C.F., and Schmid, S.E. **Handbook of Graphic Presentation**. New York: John Wiley and Sons, Inc., 1979.

Vinberg, A. **Designing a Good Graph**. San Diego: Integrated Software Systems Corporation.

Weltman, G. **Maps, A Guide to Innovative Design**. Technical Report PTR-1033S-78-1. Woodland Hills: Perceptronics, 1979.

Patrick Whitney is associate professor at the Institute of Design, Illinois Institute of Technology, where he heads the graduate and undergraduate programs in visual communication design. Prior to this he was chairman of the Design Division at the Minneapolis College of Art and Design. He was the Executive Director of The Design Foundation and a designer for RVI Corporation, a Chicago design consulting office. He was program director of the International of Graphic Design Association's Congress on design evaluation which was held at Northwestern University in 1978. He was director of the Computer Supported Design Exhibition for the 1984 SIGGRAPH conference.

Becker, E. **The Birth and Death of Meaning**. New York. The Free Press, 1971.

Bell, D. **The Coming of the Post-Industrial Society**. New York. Basic Books, Inc, 1973.

Bell, D. **The Cultural Contradictions of Capitalism**. New York. Basic Books, Inc, 1976.

Malone, T.W. **What Makes Things Fun to Learn? A Study of Intrinsically Motivating Computer Games**. Palo Alto, Xerox Palo Alto Research Center, 1980.